Glencoe McGraw-Hill

Homework Practice Workbook

Pre-Algebra

$$y = x + 2$$

McGraw Hill Glencoe

To the Student

This *Homework Practice Workbook* gives you additional problems for the concept exercises in
each lesson. The exercises are designed to aid your study of mathematics by reinforcing important
mathematical skills needed to succeed in the everyday world. The materials are organized by chapter
and lesson, with two practice worksheets for every lesson in *Glencoe Pre-Algebra*.

To the Teacher

These worksheets are the same ones found in the Chapter Resource Masters for *Glencoe Pre-Algebra*.
The answers to these worksheets are available at the end of each Chapter Resource Masters booklet.

The **McGraw·Hill** Companies

 Glencoe

Copyright © by The McGraw-Hill Companies, Inc. All rights reserved.

Except as permitted under the United States Copyright Act, no part of this publication
may be reproduced or distributed in any form or by any means, or stored in a database
or retrieval system, without prior written permission of the publisher.

Send all inquiries to:
Glencoe/McGraw-Hill
8787 Orion Place
Columbus, OH 43240

ISBN: 978-0-07-890740-1
MHID: 0-07-890740-3

Homework Practice Workbook, Pre-Algebra

Printed in the United States of America.

23 24 25 26 27 28 29 RHR   19 18 17 16 15

Contents

Lesson/Title	Page

Lesson/Title	Page

1-1   Skills Practice

Words and Expressions

Write a numerical expression for each verbal phrase.

1. the difference of seventeen and three 14

2. eleven more than six 17

3. the sum of eight, twenty, and thirty-five 63

4. the quotient of forty and eight 5

5. one hundred decreased by twenty-five 75

6. three more than one dozen 15

7. the product of twenty and thirty 600

8. five less than fifty 45

Evaluate each expression.

9. $2 - 3 \cdot 0 = -1$

10. $25 \div 5 - 4$

11. $5 + 2 - 3 = 4$

12. $2 \cdot 5 + 6$

13. $9 \div 3 \cdot 2 + 1$

14. $5 + 2 \cdot 8 + 2 - 5$

15. $4 + 2 \cdot 8$

16. $30 - 12 \cdot 2$

17. $5 + 2 \cdot 3 + 4$

18. $10 - 2 \cdot 4 - 1$

19. $15 - 10 \div 2$

20. $25 - 6 \cdot 4 + 9$

21. $(14 + 6) \div 5 = 4$

22. $100 + 50 \div 10$

23. $14 - (4 \cdot 2) = 6$

24. $(3 + 4) \cdot (5 + 3)$

25. $6(4 + 5)$

26. $\dfrac{(8 \cdot 9)}{(3 \cdot 4)}$

27. $(2 + 3) \cdot 5 + 1$

28. $24 - 24 \div 8$

29. $56 \div (3 + 4)$

30. $2[(4 + 5) \cdot 3]$

1-1 Practice

Words and Expressions

Write a numerical expression for each verbal phrase.

1. thirty-one increased by fourteen

2. the difference of sixteen and nine

3. the sum of seven, four, and eighteen

4. three times forty

5. the quotient of eighty-one and three

6. four more than the product of seven and eight

7. the cost of three slices of pizza at $2 each

8. the number of days in six weeks

Evaluate each expression.

9. $4 + 2 \cdot 8$

10. $30 - 12 \cdot 2$

11. $6(6 \div 2) \cdot 9$

12. $6(6) \div 2 \cdot 9$

13. $6(6) \div (2 \cdot 9)$

14. $6(6 \div 2 \cdot 9)$

15. $12 - 2 \cdot 5 + 3$

16. $(4 + 5) \cdot (4 + 5)$

17. $100 \div (16 + 9) \cdot 6$

18. $25 + 30 \div 6 \cdot 5$

19. $16 - 49 \div 7 \cdot 2$

20. $(2 \cdot 11 + 1) - (3 \cdot 6 + 5)$

21. $\dfrac{4(10 + 2)}{2(24 \div 3)}$

22. $2 + 4 \cdot 6 - 3 \cdot 5 + 6 \cdot 2$

23. $(8 + 4) \cdot (6 - 3)$

24. $\dfrac{2(6 + 4)}{2(8 - 6)}$

25. $4(8 + 2 \cdot 5 - 6)$

26. $2(105 \div 15 - 6)$

27. $14 \div 2 \cdot 5 + 3$

28. $4(4 + 5) \div 3(10 - 7)$

29. **BOWLING** Alicia rented bowling shoes for $3 and played 4 games at $2 each. Write and evaluate an expression for the total cost of bowling.

30. **TICKETS** Adult tickets for a movie cost $6 and children's tickets cost $3. If two adults and three children go to the movies, how much will they pay?

1-2 Skills Practice

Variables and Expressions

ALGEBRA Translate each phrase into an algebraic expression.

1. two inches shorter than Kathryn's height

2. the quotient of some number and thirteen

3. some number added to seventeen

4. six centimeters shorter than the length of the pencil

5. three pounds lighter than Adlai's weight

6. the difference of some number and eighteen

7. three dollars more than the cost of a ticket

8. eight more than the product of a number and four

9. half as many pieces of candy

10. twice as long as the length of the string

ALGEBRA Evaluate each expression if $x = 4$, $y = 6$, and $z = 3$.

11. $x + y + z$

12. $3x + y$

13. $x - z$

14. $x + y - 3z$

15. $15z$

16. $3(x + z)$

17. $xz \div y$

18. $yz - x$

ALGEBRA Evaluate each expression if $a = 7$, $b = 9$, $c = 2$, and $d = 5$.

19. $a + b + c$

20. $a + b - (c + d)$

21. $3a + 4d$

22. bcd

23. $(a + b) \cdot (c + d)$

24. $c(4 + d)$

25. $\dfrac{b}{a + c}$

26. $a + b - 3c$

27. $ab - cd$

28. $\dfrac{bc}{a - d}$

1-2 Practice

Variables and Expressions

ALGEBRA Translate each phrase into an algebraic expression.

1. six times a number minus eleven

2. the product of eight hundred and a number

3. the quotient of thirty and the product of ten times a number

4. five times the sum of three and some number

5. half the distance to the school

ALGEBRA Evaluate each expression if $x = 12$, $y = 20$, and $z = 4$.

6. $x + y + z$

7. $4x - y$

8. $3x + 2y$

9. $y - 3z$

10. $x + y \div z$

11. $yz + x$

12. $(y - x) + (y - z)$

13. $\frac{y}{z} + \frac{x}{z}$

14. $\frac{5x}{3y}$

15. $z(y - x) + 4z$

ALGEBRA Evaluate each expression if $a = 3$, $b = 6$, $c = 5$, and $d = 9$.

16. $a + b + c + d$

17. $\frac{(a + b + c)}{2}$

18. $ab + bc$

19. $6d - c \cdot c$

20. $3(a + b + c)$

21. $\frac{100}{5c}$

22. abc

23. $10(6c - 3d)$

24. $\frac{2(a + b)}{6(b - c)}$

25. $4[(d - a) + c]$

26. **RECYCLING** In order to encourage recycling, the city is offering five cents for every pound of newspapers collected, twenty-five cents per pound for cans, and ten cents per pound for glass bottles or jars.

   a. Write an expression for the total amount earned from recycling.

   b. If Chen brings in ten pounds of newspapers, eight pounds of cans, and two pounds of glass, how much will he receive?

1-3 Skills Practice

Properties

Name the property shown by each statement.

1. $9 \cdot 7 = 7 \cdot 9$

2. $37 \cdot 0 = 0$

3. $1 \cdot 87 = 87$

4. $14 + 6 = 6 + 14$

5. $3(6a) = (3 \cdot 6)a$

6. $2b + 0 = 2b$

7. $4 + (6 + 8) = (4 + 6) + 8$

8. $2 \cdot (15 \cdot 10) = 2 \cdot (10 \cdot 15)$

ALGEBRA Simplify each expression.

9. $(a \cdot 0) \cdot 6$

10. $b + (7 + 5)$

11. $4w(9)$

12. $20(t \cdot 1)$

13. $(x + 5) + 4$

14. $(6a)10$

15. $38 + (v + 12)$

16. $8(3q)$

17. $16p \cdot 0$

18. $16 + (22 + x)$

19. $8(9p)$

20. $(17 + 33) + x$

21. $3(11k)$

22. $16 + (y + 9)$

23. $m(13 \cdot 5)$

24. $17 + (n + 0)$

1-3 Practice

Properties

Name the property shown by each statement.

1. $55 + 6 = 6 + 55$

2. $6 \cdot 7 = 7 \cdot 6$

3. $(x + 3) + y = x + (3 + y)$

4. $1 \cdot mp = mp$

5. $9 + (5 + 35) = (9 + 5) + 35$

6. $67 + 0 = 67$

7. $7x \cdot 0 = 0$

8. $4(3 \cdot z) = (4 \cdot 3)z$

ALGEBRA   Simplify each expression.

9. $a(5 \cdot 7)$

10. $(24 + s) + 16$

11. $c + (17 + 8)$

12. $72g(1)$

13. $31 + (21 + p)$

14. $(e \cdot 4) \cdot 12$

15. $(m + 11) + 19$

16. $(9 \cdot b) \cdot 10$

17. $19 + (v + 8)$

18. $(28 + 12) + x$

19. $8s \cdot 0$

20. $4 \cdot (r \cdot 5)$

21. **GEOMETRY** The volume of a box is given by $V = \ell \cdot w \cdot h$ where ℓ = length, w = width, and h = height. Find the volume of a box if the length is 25 cm, width is 13 cm, and height is 4 cm.

22. **SCHOOL** In math class each assignment is worth 20 points. David got 17, 20, 19, and 13 points on his last four assignments. How many points did David score all together?

23. State whether the following statement is true or false: Multiplying any number by one produces the original number. Explain.

1-4 Skills Practice

Ordered Pairs and Relations

Graph each ordered pair on the coordinate plane.

1. $A(2, 5)$

2. $M(6, 4)$

3. $Z(1, 1)$

4. $R(3, 0)$

5. $Q(7, 8)$

6. $W(0, 6)$

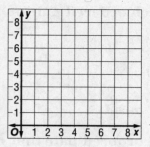

Write the ordered pair that names each point.

7. N

8. K

9. A

10. V

11. Z

12. G

13. R

14. B

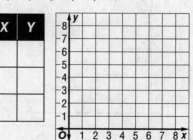

Express each relation as a table and as a graph. Then determine the domain and range.

15. $\{(3, 7), (1, 1), (6, 5), (2, 4)\}$

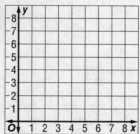

16. $\{(0, 3), (5, 7), (1, 8)\}$

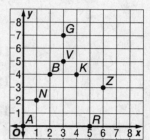

17. $\{(2, 3), (3, 2), (1, 7), (7, 1)\}$

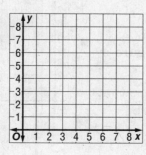

18. $\{(5, 6), (0, 2), (4, 4) (8, 3)\}$

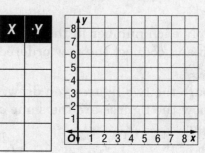

1-4 Practice

Ordered Pairs and Relations

Graph each ordered pair on the coordinate plane.

1. $Q(4, 2)$

2. $V(3, 7)$

3. $T(0, 3)$

4. $B(8, 6)$

5. $R(5, 0)$

6. $L(4, 4)$

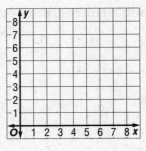

Write the ordered pair that names each point.

7. J

8. X

9. R

10. B

11. K

12. H

13. D

14. N

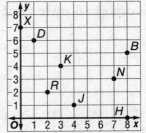

Express each relation as a table and as a graph. Then determine the domain and range.

15. $\{(3, 7), (1, 1), (6, 5), (2, 4)\}$

 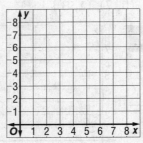

16. $\{(0, 2), (4, 6), (3, 7)\}$

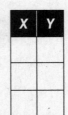 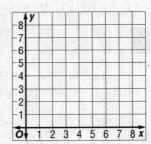

17. **GEOMETRY** Graph $(2, 1)$, $(2, 4)$, and $(5, 1)$ on the coordinate system.

   a. Connect the points with line segments. What figure is formed?

   b. Multiply each number in the set of ordered pairs by 2. Graph and connect the new ordered pairs. What figure is formed?

   c. Compare the two figures you drew. Write a sentence that tells how the figures are the same and how they are different.

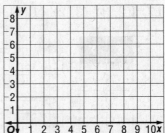

1-5 Skills Practice

Words, Equations, Tables, and Graphs

Copy and complete each function table. Then state the domain and range of the function.

1. A phone call costs $3 a minute.

Input (x)	Rule:	Output (y)
1		
5		
10		
20		

2. Jared has 4 less than 3 times the number of trophies that Zach has.

Input (x)	Rule:	Output (y)
2		
4		
6		
8		

3. The cost for a class trip is $5 per student plus $100 for the bus.

Input (x)	Rule:	Output (y)
18		
22		
24		
30		

4. A child's admission is $4 more than half an adult's admission.

Input (x)	Rule:	Output (y)
20		
26		
30		
42		

5. MULTIPLE REPRESENTATIONS There are 12 inches in 1 foot.

   a. ALGEBRAIC Write an equation to find the number of inches in any number of feet.

   b. TABULAR Make a function table to find the number of inches in 4, 6, 8, and 10 feet.

   c. GRAPHICAL Graph the ordered pairs for the function.

Input (x)	Rule:	Output (y)

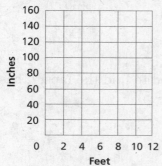

1-5 Practice

Words, Equations, Tables, and Graphs

Copy and complete each function table. Then state the domain and range of the function.

1. Each copy of a book costs $18 and shipping is $9 per order.

Input (x)	Rule:	Output (y)
2		
3		
4		
5		

2. The number of girls at a camp is 17 less than twice the number of boys.

Input (x)	Rule:	Output (y)
40		
58		
82		
100		

3. Tom's height is 10 inches more than one third of his older sister's height.

Input (x)	Rule:	Output (y)
66		
57		
51		
42		

4. The charge for a hotel room is $75 per night plus a $15 booking fee.

Input (x)	Rule:	Output (y)
2		
4		
8		
14		

5. WALKING Carly walked laps in a charity walk-a-thon. The graph shows the number of laps walked over 80 minutes.

a. TABULAR Make a function table showing the input, minutes, and the output, laps walked.

Input (x)	Output (y)

b. ALGEBRAIC Can you write one equation that can be used to find the laps, l, based on the minutes, m? Explain.

c. Is the relation a function? Explain.

Copyright © Glencoe/McGraw-Hill, a division of The McGraw-Hill Companies, Inc.

1-6 Skills Practice

Scatter Plots

Tell whether each scatter plot shows a *positive*, *negative*, or *no* relationship.

1.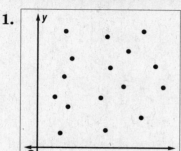

2.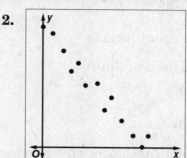

3.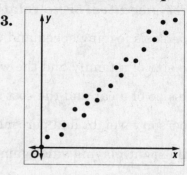

4. Draw a scatter plot with six ordered pairs that shows a positive relationship. Explain your reasoning.

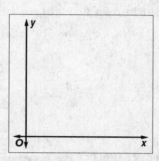

For Exercises 5–8, use the following information.

SCIENCE Scientists have determined that there may be a relationship between temperature and the number of chirps produced by crickets. The table gives the temperature and the number of chirps per minute for several cricket samples.

5. Make a scatter plot of the data.

6. Does there appear to be a relationship between temperature and chirps? Explain.

Temperature (°F)	Chirps/min
71	138
68	97
75	152
80	158
60	81
75	155
84	165

7. Suppose the outside temperature is 65°. About how many chirps per minute would you expect from a cricket?

8. Suppose the outside temperature is 55°. About how many chirps per minute would you expect from a cricket?

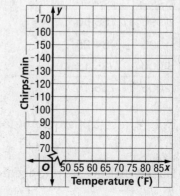

1-6 Practice

Scatter Plots

Determine whether a scatter plot of the data for the following might show a *positive*, *negative*, or *no* relationship.

1. a person's jogging speed and time spent jogging

2. the size of a family and the weekly grocery bill

3. the size of a car and the cost

4. a person's weight and percent body fat

5. time spent playing video games and time spent on outdoor activities

6. Draw a scatter plot with ten ordered pairs that shows a negative relationship.

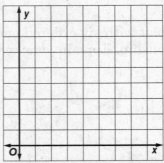

EMPLOYMENT **For Exercises 7–9, use the table below, which shows the federal minimum wage rates from 1950 to 2000.**

Year	Minimum Wage
1950	$0.75
1955	$0.75
1960	$1.00
1965	$1.25
1970	$1.60
1975	$2.10
1980	$3.10
1985	$3.35
1990	$3.80
1995	$4.25
2000	$5.15
2005	$5.15

7. Make a scatter plot of the data.

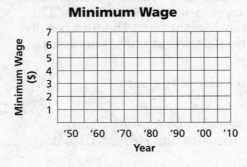

8. Does there appear to be a relationship between year and minimum wage?

9. Based on the graph, predict what the minimum wage will be for the year 2010.

2-1 Skills Practice

Integers and Absolute Value

Write an integer for each situation. Then graph on a number line.

1. a bank deposit of $200

2. a loss of 10 yards

3. 450 feet above sea level

4. 7°F below normal

Replace each ● with <, >, or = to make a true sentence.

5. 1 ● 0

6. −3 ● 0

7. 0 ● −1

8. 0 ● 9

9. −7 ● −7

10. 2 ● −2

11. −2 ● 8

12. −4 ● 4

13. 5 ● 5

14. 0 ● −6

15. 4 ● 10

16. 6 ● −6

17. 3 ● 7

18. −1 ● −2

19. 3 ● 4

20. −3 ● −4

Evaluate each expression.

21. $|1|$

22. $|-10|$

23. $|-8|$

24. $|10|$

25. $|4| + |-4|$

26. $|9| - |-5|$

27. $0 + |-1|$

28. $|-6| + |-5|$

29. $|-8| - |-8|$

30. $|12| + |-3|$

31. $|-15| - |6|$

32. $|-13| + |-7|$

ALGEBRA Evaluate each expression if $a = -3$, $b = 0$, and $c = 1$.

33. $|a| - b$

34. $|c| + 2$

35. $9 - |a|$

36. $25 - |b|$

37. $10 - |b|$

38. $|-8| + |a|$

2-1 Practice

Integers and Absolute Value

Write an integer for each situation. Then graph on a number line.

1. an elevator ascends 4 floors

2. to be at par

3. 11°F below zero

4. a profit of $52 on a sale

Replace each ● with <, >, or = to make a true sentence.

5. $0 ● -5$

6. $10 ● -10$

7. $-8 ● 3$

8. $11 ● 11$

9. $-18 ● -18$

10. $-18 ● 18$

11. $18 ● -18$

12. $18 ● 18$

13. $-120 ● -95$

14. $35 ● -12$

15. $-35 ● 12$

16. $41 ● 17$

Evaluate each expression.

17. $|-7|$

18. $|14|$

19. $|-11|$

20. $|-9| - |6|$

21. $|-18| - |-8|$

22. $|-12| + |1|$

23. $|8 - 4|$

24. $|23| - |18|$

25. $|-16| + |-22|$

ALGEBRA Evaluate each expression if $a = -3$, $b = 0$, and $c = 1$.

26. $|a| - |c|$

27. $|a| + |c|$

28. $|ab| + c$

29. $5 - |ac|$

30. $c + |-5|$

31. $c + |5|$

32. WEATHER At 6:15 A.M. the temperature was -8°F. At 12:15 P.M. the temperature was -12°F. At 6:16 P.M. the temperature was -10°F. Order the temperatures from least to greatest.

2-2 Skills Practice

Adding Integers

Find each sum.

1. $-7 + (-5)$

2. $10 + 9$

3. $-12 + (-5)$

4. $-13 + (-3)$

5. $-10 + 12$

6. $-7 + 8$

7. $-11 + (-6)$

8. $0 + (-21)$

9. $72 + (-10)$

10. $72 + 10$

11. $-13 + (-11)$

12. $-52 + 52$

13. $-6 + (-12)$

14. $14 + (-8)$

15. $-17 + (-2)$

16. $50 + (-8)$

17. $-22 + 4$

18. $14 + 8$

19. $-21 + (-9)$

20. $15 + (-5)$

21. $9 + 10$

22. $-12 + (-15)$

23. $-13 + 6$

24. $-1 + (-18)$

25. $0 + 31$

26. $-45 + (-15)$

27. $-6 + 20$

28. $24 + (-11)$

29. $7 + (-14)$

30. $-34 + (-10)$

31. $-8 + (-25)$

32. $-31 + 25$

33. $4 + 5 + (-4)$

34. $-4 + (-5) + 6$

35. $-3 + 8 + (-9)$

36. $-6 + (-2) + (-1)$

37. $10 + (-5) + 6$

38. $-8 + 8 + (-10)$

39. $0 + (-8) + 22$

40. $-31 + 19 + (-19)$

41. $32 + (-4) + (-9)$

2-2 Practice

Adding Integers

Find each sum.

1. $-19 + (-7)$ 2. $-29 + 30$ 3. $-32 + 9$ 4. $10 + 37$

5. $34 + 22$ 6. $-16 + (-28)$ 7. $-4 + (-50)$ 8. $-12 + (-63)$

9. $26 + (-9)$ 10. $-17 + (-23)$ 11. $12 + (-22)$ 12. $18 + (-56)$

13. $-36 + (-36)$ 14. $-54 + 45$ 15. $-34 + 17$ 16. $-16 + (-24)$

17. $70 + (-108)$ 18. $-52 + 36$ 19. $-71 + (-86)$ 20. $-39 + (-40)$

21. $25 + 18 + (-25)$ 22. $-65 + (-2) + 9$ 23. $80 + 15 + (-26)$

24. $-5 + 4 + (-27)$ 25. $-29 + 29 + 44$ 26. $-1 + (-8) + (-49)$

27. $-16 + (-56) + (-90)$ 28. $-18 + 13 + (-35)$ 29. $10 + (-34) + 34$

30. $30 + (-9) + 1$ 31. $-24 + 7 + (-7)$ 32. $51 + (-21) + (-12)$

33. **TEMPERATURE** At 4:00 A.M., the outside temperature was −28°F. By 4:00 P.M. it rose 38 degrees. What was the temperature at 4:00 P.M.?

34. **HEALTH** Three friends decided to exercise together four times a week to lose fat and increase muscle mass. While all three were healthier after six weeks, one had lost 5 pounds, another had gained 3 pounds, and one had lost 4 pounds. What was the total number of pounds gained or lost by the three friends?

35. **ROLLER COASTERS** The latest thrill ride at a popular theme park takes roller coaster fans on an exciting ride. In the first 20 seconds, it carries its passengers up a 100-meter hill, plunges them 72 meters down, and quickly takes them back up a 48-meter rise. How much higher or lower from the start of the ride are they after these 20 seconds?

2-3 Skills Practice

Subtracting Integers

Find each difference.

1. $-2 - (-8)$ **2.** $4 - (-11)$ **3.** $-7 - 6$ **4.** $15 - 2$

5. $-7 - (-1)$ **6.** $1 - 9$ **7.** $-5 - (-3)$ **8.** $6 - (-5)$

9. $-4 - (-10)$ **10.** $4 - 6$ **11.** $0 - (-15)$ **12.** $-16 - (-10)$

13. $0 - 16$ **14.** $11 - (-9)$ **15.** $-9 - 1$ **16.** $-1 - (-8)$

17. $1 - (-2)$ **18.** $-2 - (-19)$ **19.** $13 - 17$ **20.** $20 - (-15)$

21. $-10 - (-21)$ **22.** $4 - 22$ **23.** $-8 - 16$ **24.** $12 - (-9)$

ALGEBRA **Evaluate each expression if $a = -9$, $b = 4$, and $c = -5$.**

25. $a - 8$ **26.** $10 - c$ **27.** $11 - b$ **28.** $15 - a$

29. $-8 - b$ **30.** $c - 1$ **31.** $-32 - a$ **32.** $b - 25$

33. $c - (-14)$ **34.** $-33 - a$ **35.** $14 - c$ **36.** $b - c$

37. $a - c$ **38.** $b - a$ **39.** $c - b$ **40.** $c - a$

41. $a - b$ **42.** $a + b - c$ **43.** $b + 15 + a$ **44.** $a - (-b) + c$

2-3 Practice

Subtracting Integers

Find each difference.

1. $-26 - (-30)$ **2.** $25 - 32$ **3.** $-18 - 54$ **4.** $59 - (-19)$

5. $-41 - (-19)$ **6.** $-20 - 13$ **7.** $31 - (-56)$ **8.** $15 - (-40)$

9. $-32 - 28$ **10.** $10 - (-23)$ **11.** $-14 - 64$ **12.** $-12 - (-36)$

13. $-81 - 4$ **14.** $9 - 30$ **15.** $-44 - (-21)$ **16.** $140 - (-9)$

Evaluate each expression if $a = -11$, $b = 8$, and $c = -6$.

17. $a - 17$ **18.** $10 - b$ **19.** $-30 - c$ **20.** $b - a$

21. $a - b$ **22.** $c - b$ **23.** $b - c + a$ **24.** $b - c - a$

25. $c - a - b$ **26.** $b + a - c$ **27.** $b + c - a$ **28.** $c - a + b$

29. $a - b + c$ **30.** $b - a + c$ **31.** $a - b - c$ **32.** $c + b - a$

33. $c - b + a$ **34.** $a + b - c$ **35.** $16 + a + c$ **36.** $a - b + 14$

37. ELEVATORS Linda entered an elevator on floor 9. She rode down 8 floors. Then she rode up 11 floors and got off. What floor was she on when she left the elevator?

38. INVESTMENTS The NASDAQ lost 36 points on a Monday, but rebounded the next day, gaining 24 points. What was the total change in points?

39. OFFICE BUILDINGS Randi takes the stairs at work whenever possible instead of the elevator. She must climb up 51 steps from her office to get to the accounting department. The human resources department is 34 steps below her office. How many steps are there between human resources and accounting?

2-4 Skills Practice

Multiplying Integers

Find each product.

1. $-2(8)$

2. $-4(-4)$

3. $6(-2)$

4. $-7(-3)$

5. $12(1)$

6. $0(-2)$

7. $-11(5)$

8. $-9(-3)$

9. $-13(0)$

10. $-1(-1)$

11. $-1(1)$

12. $1(-1)$

13. $-5(20)$

14. $16(-2)$

15. $18(-3)$

16. $-5(-5)$

17. $8(6)(-2)$

18. $-1(50)(-1)$

19. $6(0)(-2)$

20. $(-3)(-2)(-1)$

21. $-4(5)(-3)$

22. $10(-3)(2)$

23. $-9(8)(1)$

24. $-1(-1)(-1)$

ALGEBRA Simplify each expression.

25. $-2 \cdot 3x$

26. $-4 \cdot 5y$

27. $9 \cdot (-2z)$

28. $-5 \cdot (-6a)$

29. $8t \cdot (-3)$

30. $2n \cdot (-1)$

31. $-5 \cdot 2w$

32. $8c \cdot (-2)$

33. $-3c \cdot (-5d)$

34. $4r \cdot 7s$

35. $-3x \cdot (-z)$

36. $-4ab \cdot (-6)$

37. $(-3)(4)(-x)$

38. $-3(5)(-y)$

39. $(-6)(-2)(8r)$

40. $-5(0)(-xy)$

ALGEBRA Evaluate each expression if $x = -5$ and $y = -6$.

41. $3y$

42. $-8x$

43. $-4y$

44. $12x$

45. xy

46. $-xy$

47. $-6xy$

48. $4xy$

2-4 Practice

Multiplying Integers

Find each product.

1. 8(16)

2. −4(17)

3. −1(−40)

4. −5(−7)

5. 0(−54)

6. 29(−2)

7. −20(−20)

8. −31(−4)

9. −2(−15)(−6)

10. 3(−5)(−8)

11. −10(17)(−2)

12. −2(−2)(−2)

13. 12(10)(5)

14. −50(−21)(2)

15. −8(−13)(−25)

16. −5(16)(4)

ALGEBRA Simplify each expression.

17. $-6r \cdot (12s)$

18. $-15 \cdot (9v)$

19. $2ab \cdot (-25)$

20. $-27y \cdot (-z)$

21. $-60m(-2)(-3n)$

22. $-9u(-4)(-w)$

23. $29g(0)(-15)$

24. $-b(-12)(11)$

25. $19h(-1)(-2s)$

26. $-h(-jk)$

27. $(-1)(-a)(-bc)$

28. $(-1)(-fg)(-xy)$

ALGEBRA Evaluate each expression if $a = -1$, $b = -6$, and $c = 5$.

29. $-11a$

30. $4ab$

31. $-8bc$

32. $-10ac$

33. $15ab$

34. $12ac$

35. abc

36. $-abc$

37. $-11a(-bc)$

38. $4ab(-8c)$

39. $9a(-2b)(5c)$

40. $-3a(-2b)(-c)$

41. REAL ESTATE In Montyville, the value of homes has experienced an annual change of −2 percent. If the rate continues, what will be the change over 10 years?

42. RETAIL The Good Food n' More grocery store loses an average of $210 a day due to breakage, shoplifting, and food expiration. How much money does the store lose on average per 7-day week?

2-5 Skills Practice

Dividing Integers

Find each quotient.

1. $16 \div 4$ **2.** $-27 \div 3$ **3.** $25 \div (-5)$ **4.** $63 \div (-9)$

5. $-15 \div (-3)$ **6.** $14 \div (-7)$ **7.** $-124 \div 4$ **8.** $60 \div 15$

9. $28 \div (-4)$ **10.** $-56 \div (-8)$ **11.** $72 \div 8$ **12.** $-21 \div (-7)$

13. $\dfrac{-32}{4}$ **14.** $\dfrac{45}{9}$ **15.** $\dfrac{-45}{3}$ **16.** $\dfrac{-25}{-5}$

17. $\dfrac{35}{-7}$ **18.** $\dfrac{-63}{-7}$ **19.** $\dfrac{-144}{12}$ **20.** $\dfrac{48}{-6}$

21. What is -54 divided by 9?

22. Divide -27 by -3.

23. Divide 144 by -12.

24. What is -65 divided by -13?

ALGEBRA Evaluate each expression if $x = -8$ and $y = -12$.

25. $x \div 2$ **26.** $x \div (-4)$ **27.** $36 \div y$ **28.** $0 \div y$

29. $-60 \div y$ **30.** $56 \div x$ **31.** $8 \div x$ **32.** $-108 \div y$

33. $\dfrac{x}{-2}$ **34.** $\dfrac{y}{3}$ **35.** $\dfrac{0}{x}$ **36.** $\dfrac{-112}{x}$

37. $\dfrac{y}{-6}$ **38.** $\dfrac{x}{4}$ **39.** $\dfrac{-144}{y}$ **40.** $\dfrac{-136}{x}$

Find the average (mean) of each group of numbers.

41. 3, 12, 6   **42.** $-8, -1, -3$ **43.** $-8, 15, 5, 8$ **44.** $-3, -10, 2, -4, 0$

45. $-10, -7, 7, 10$ **46.** 12, 24, 9, 15, 18, 20, 16, 14   **47.** $-4, -11, -6, 1, 8, -12$

2-5 Practice

Dividing Integers

Find each quotient.

1. $-44 \div 4$ **2.** $0 \div (-5)$ **3.** $-21 \div 21$ **4.** $32 \div 8$

5. $-17 \div (-17)$ **6.** $-49 \div 7$ **7.** $80 \div (-4)$ **8.** $-64 \div (-8)$

9. $\dfrac{72}{-9}$ **10.** $\dfrac{-100}{-5}$ **11.** $\dfrac{-90}{6}$ **12.** $\dfrac{360}{12}$

13. $\dfrac{-400}{-25}$ **14.** $\dfrac{-525}{5}$ **15.** $\dfrac{84}{-6}$ **16.** $\dfrac{215}{5}$

ALGEBRA Evaluate each expression if $a = -2$, $b = 5$, and $c = -4$.

17. $-35 \div b$ **18.** $54 \div a$ **19.** $-56 \div c$ **20.** $205 \div b$

21. $\dfrac{c}{-2}$ **22.** $\dfrac{b}{5}$ **23.** $\dfrac{2}{a}$ **24.** $\dfrac{-4}{c}$

25. $\dfrac{-28}{c}$ **26.** $\dfrac{ac}{-8}$ **27.** $\dfrac{bc}{a}$ **28.** $\dfrac{250}{ab}$

Find the average (mean) of each group of numbers.

29. $23, 20, 27, 18$ **30.** $-8, 9, 4, 0, 2, -1$ **31.** $17, 21, 4$

32. $-20, -15, -12, -1, 1, 12, 15, 20$ **33.** $-7, -3, -9, 0, 21, -2, -14$

34. TESTS Miranda earned scores of 84, 91, 95, 78, and 92 on her math tests. Find her average (mean) score.

35. TEMPERATURE At noon on Friday, the temperature was 0°F. Six hours later the temperature was −18°F. On average, what was the temperature change per hour?

36. BUSINESS The architecture firm of Stuart and Maxwell, Ltd., had monthly profits of $1200, $755, −$450, $210, and −$640 over 5 months. What was the average profit for those months?

2-6 Skills Practice

Graphing in Four Quadrants

Name the ordered pair for each point graphed at the right.

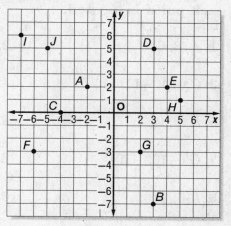

1. A

2. B

3. C

4. D

5. E

6. F

7. G

8. H

9. I

10. J

Graph and label each point on the coordinate plane. Name the quadrant in which each point is located.

11. K (1, 0)

12. L (0, 2)

13. M (−2, 4)

14. N (−5, −4)

15. P (6, −2)

16. Q (7, −6)

17. R (−3, −4)

18. S (4, −7)

19. T (3, 6)

20. U (−7, 4)

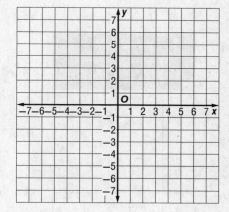

21. **ALGEBRA** Make a table of values and graph six sets of ordered pairs for the equation $y = x - 4$. Describe the graph.

y = x − 4		
x	y	(x, y)

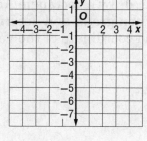

2-6 Practice

Graphing in Four Quadrants

Graph and label each point on the coordinate plane. Name the quadrant in which each point is located.

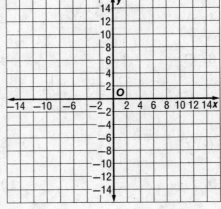

1. A (8, 6)

2. B (−8, 6)

3. C (−4, −11)

4. D (3, −6)

5. E (9, 0)

6. F (−4, 1)

7. G (−10, −10)

8. H (0, −8)

9. I (6, −2)

10. J (2, 13)

11. ALGEBRA Make a table of values and graph six sets of ordered pairs for the equation $y = 5 − x$. Describe the graph.

$y = 5 − x$		
x	**y**	**(x, y)**

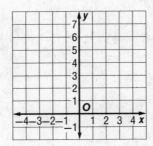

12. GEOMETRY On the coordinate plane, draw a rectangle $ABCD$ with vertices at $A(1, 4)$, $B(5, 4)$, $C(5, 1)$, and $D(1, 1)$. Then graph and describe the new rectangle formed when you subtract 3 from each coordinate of the vertices in rectangle $ABCD$.

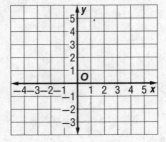

2-7   Skills Practice

Translations and Reflections on the Coordinate Plane

For Exercises 1 and 2, use the coordinate plane below. Triangle *PQR* is shown.

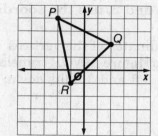

1. Find the coordinates of the vertices of the image of △*PQR* translated 3 units to the left and 4 units down.

2. Find the coordinates of the vertices of the image of △*PQR* translated 2 units to the right and 5 units down.

For Exercises 3 and 4, use the coordinate plane below. Figure *ABCD* is shown.

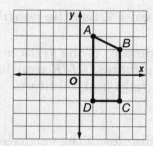

3. Find the coordinates of the vertices of the image of figure *ABCD* translated 1 unit to the right and 6 units down.

4. Find the coordinates of the vertices of the image of figure *ABCD* translated 4 units to the left and 2 units up.

5. The vertices of figure *HJKL* are $H(3, 1)$, $J(5, -2)$, $K(1, -4)$, and $L(1, 0)$. Graph the figure and its image after a reflection over the *y*-axis.

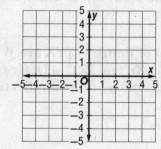

6. The vertices of figure *STUV* are $S(-3, 2)$, $T(-2, 4)$, $U(3, 3)$, and $V(2, 1)$. Graph the figure and its image after a reflection over the *x*-axis.

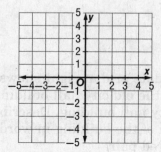

2-7 Practice

Translations and Reflections on the Coordinate Plane

For Exercises 1 and 2, use the coordinate plane below. Figure *MNOP* is shown.

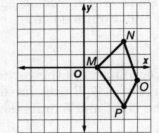

1. Find the coordinates of the vertices of the image of figure *MNOP* translated 6 units to the left and 3 units down.

2. Find the coordinates of the vertices of the image of figure *MNOP* translated 4 units to the left and 4 units up.

For Exercises 3 and 4, use the coordinate plane below. Figure *GHJK* is shown.

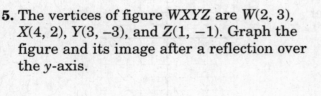

3. Find the coordinates of the vertices of the image of figure *GHJK* translated 3 units to the right and 1 unit up.

4. Find the coordinates of the vertices of the image of figure *GHJK* translated 5 units to the right and 2 units down.

5. The vertices of figure *WXYZ* are $W(2, 3)$, $X(4, 2)$, $Y(3, -3)$, and $Z(1, -1)$. Graph the figure and its image after a reflection over the y-axis.

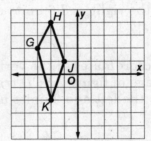

6. The coordinate grid at right shows the location of the old city hall building and the new city hall building. Describe the translation of the building in words and as an ordered pair.

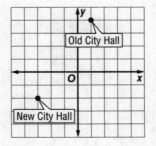

3-1 Skills Practice

Fractions and Decimals

Write each fraction as a decimal. Use a bar to show a repeating decimal.

1. $\frac{3}{4}$

2. $\frac{2}{5}$

3. $\frac{5}{10}$

4. $\frac{5}{5}$

5. $\frac{13}{100}$

6. $\frac{4}{5}$

7. $\frac{7}{10}$

8. $\frac{7}{8}$

9. $\frac{1}{4}$

10. $\frac{7}{50}$

11. $-\frac{3}{10}$

12. $\frac{1}{11}$

13. $\frac{1}{8}$

14. $\frac{1}{12}$

15. $\frac{7}{30}$

16. $\frac{1}{15}$

17. $\frac{7}{11}$

18. $-\frac{5}{9}$

19. $-\frac{3}{5}$

20. $\frac{1}{6}$

Replace each ● with $<$, $>$, or $=$ to make a true sentence.

21. $\frac{1}{8}$ ● 0.12

22. $\frac{2}{3}$ ● 0.7

23. $-\frac{3}{10}$ ● -0.3

24. 0.395 ● $\frac{2}{5}$

25. 0.1 ● $\frac{1}{11}$

26. $0.\overline{16}$ ● $\frac{1}{6}$

27. $\frac{3}{5}$ ● $\frac{3}{4}$

28. $-\frac{1}{4}$ ● -0.25

29. Order $\frac{9}{11}$, 0.99, and $\frac{9}{10}$ from least to greatest.

30. Order 0.5, $\frac{4}{9}$, and $\frac{2}{5}$ from least to greatest.

3-1 Practice

Fractions and Decimals

Write each fraction as a decimal. Use a bar to show a repeating decimal.

1. $\dfrac{3}{5}$

2. $\dfrac{1}{8}$

3. $\dfrac{9}{11}$

4. $-\dfrac{3}{16}$

5. $\dfrac{3}{40}$

6. $\dfrac{8}{11}$

7. $\dfrac{5}{12}$

8. $\dfrac{1}{3}$

9. $\dfrac{7}{9}$

10. $-\dfrac{11}{15}$

11. $-\dfrac{12}{16}$

12. $\dfrac{13}{60}$

13. $\dfrac{1}{45}$

14. $-\dfrac{5}{24}$

15. $\dfrac{13}{20}$

16. $\dfrac{17}{18}$

17. $-\dfrac{1}{4}$

18. $\dfrac{5}{11}$

19. $-\dfrac{2}{3}$

20. $\dfrac{7}{8}$

Replace each ● with $<$, $>$, or $=$ to make a true sentence.

21. $-\dfrac{13}{2}$ ● -6.4

22. $\dfrac{6}{7}$ ● $\dfrac{5}{6}$

23. -0.75 ● $-\dfrac{15}{20}$

24. $-\dfrac{3}{8}$ ● -0.40

25. $\dfrac{7}{8}$ ● $\dfrac{8}{9}$

26. $-\dfrac{33}{100}$ ● $-0.\overline{3}$

27. Order $\dfrac{4}{9}$, $\dfrac{444}{1000}$, and 0.4 from least to greatest.

28. Order $-\dfrac{8}{9}$, $-\dfrac{8}{10}$, and $-0.\overline{80}$ from least to greatest.

29. **OPINION** In a school survey, 787 out of 1000 students preferred hip-hop music to techno. Is this figure more or less than $\dfrac{7}{9}$ of those surveyed? Explain.

3-2 Skills Practice

Rational Numbers

Write each number as a fraction.

1. 13

2. $1\frac{1}{4}$

3. 57

4. -25

5. $-3\frac{4}{5}$

6. $6\frac{5}{8}$

7. -1

8. $2\frac{2}{9}$

Write each decimal as a fraction or mixed number in simplest form.

9. 0.6

10. 0.25

11. $0.\overline{4}$

12. $-1.\overline{1}$

13. 0.11

14. 2.8

15. 7.03

16. -2.12

17. $3.\overline{2}$

18. 1.125

19. 8.65

20. 16.7

21. 0.16

22. 4.06

23. $-5.\overline{8}$

24. $0.\overline{24}$

25. Write 85 hundredths as a fraction in simplest form.

26. Write 9 and 250 thousandths as a mixed number in simplest form.

Identify all sets to which each number belongs (W = whole numbers, I = integers, Q = rational numbers).

27. 16

28. -2.54

29. $\frac{9}{3}$

30. $0.\overline{95}$

31. -4

32. 2.2020020002. . .

3-2 Practice

Rational Numbers

Write each number as a fraction.

1. 29

2. 0

3. $3\frac{7}{8}$

4. -47

5. $-5\frac{6}{7}$

6. $4\frac{3}{20}$

7. $-7\frac{2}{15}$

8. $10\frac{2}{9}$

Write each decimal as a fraction or mixed number in simplest form.

9. 0.32

10. 0.42

11. $0.\overline{8}$

12. $-6.\overline{3}$

13. 0.91

14. 17.875

15. $-0.666\ldots$

16. 0.07

17. $9.\overline{7}$

18. 7.75

19. 0.525

20. -8.26

21. $6.\overline{5}$

22. -4.12

23. 13.006

24. $3.\overline{34}$

Identify all sets to which each number belongs (W = whole numbers, I = integers, Q = rational numbers).

25. 15

26. $-3.\overline{8}$

27. -5.075

28. $\frac{50}{25}$

29. π

30. $-\frac{4}{2}$

31. **BOTANY** The smallest flowering plant is the flowering aquatic duckweed found in Australia. It is 0.0236 inch long and 0.0129 inch wide. Write these dimensions as fractions in simplest form.

3-3   Skills Practice

Multiplying Rational Numbers

Find each product. Write in simplest form.

1. $\frac{1}{3} \cdot \left(-\frac{1}{4}\right)$

2. $-\frac{2}{5} \cdot \frac{6}{7}$

3. $\frac{2}{7} \cdot \frac{3}{11}$

4. $\frac{3}{13} \cdot \frac{2}{5}$

5. $\frac{2}{9} \cdot \frac{3}{5}$

6. $\frac{3}{11} \cdot \frac{5}{9}$

7. $-\frac{1}{4} \cdot \frac{4}{9}$

8. $\frac{3}{5} \cdot \frac{15}{18}$

9. $\frac{3}{4} \cdot \frac{2}{5}$

10. $-\frac{1}{6} \cdot \left(-\frac{4}{7}\right)$

11. $\frac{5}{14} \cdot \left(-\frac{7}{9}\right)$

12. $-\frac{2}{3} \cdot \frac{9}{10}$

13. $\frac{5}{16} \cdot 4$

14. $5\frac{1}{2} \cdot \frac{2}{11}$

15. $-3 \cdot \left(-\frac{8}{9}\right)$

16. $-\frac{3}{5} \cdot 6\frac{2}{3}$

17. $-12\frac{2}{3} \cdot 7\frac{1}{2}$

18. $-\frac{5}{36} \cdot \left(-\frac{9}{25}\right)$

19. $8\frac{4}{5} \cdot 2\frac{5}{10}$

20. $3\frac{1}{3} \cdot 9\frac{3}{4}$

21. $-6\frac{2}{5} \cdot \left(-2\frac{2}{9}\right)$

22. $\frac{7}{45} \cdot \frac{9}{42}$

ALGEBRA Evaluate each expression if $a = \frac{9}{12}$, $b = -2\frac{1}{4}$, and $c = \frac{2}{5}$. Write the product in simplest form.

23. ab

24. $-2b$

25. $\frac{5}{8}ac$

26. $3bc$

27. $-5\frac{1}{2}ab$

28. $\frac{1}{3}abc$

3-3 Practice

Multiplying Rational Numbers

Find each product. Write in simplest form.

1. $\dfrac{3}{4} \cdot \dfrac{2}{3}$

2. $\dfrac{3}{7} \cdot \dfrac{21}{39}$

3. $-\dfrac{3}{4} \cdot \dfrac{10}{27}$

4. $\dfrac{11}{14} \cdot \dfrac{7}{33}$

5. $-\dfrac{18}{24} \cdot \dfrac{3}{4}$

6. $\dfrac{9}{10} \cdot \dfrac{20}{21}$

7. $-50 \cdot \dfrac{3}{1000}$

8. $\dfrac{16}{17} \cdot \left(-\dfrac{5}{8}\right)$

9. $-\dfrac{1}{2} \cdot \left(-\dfrac{20}{27}\right)$

10. $-\dfrac{14}{15} \cdot \left(-\dfrac{10}{28}\right)$

11. $4\dfrac{4}{7} \cdot 9\dfrac{1}{3}$

12. $-2\dfrac{14}{25} \cdot \dfrac{3}{8}$

13. $4\dfrac{1}{8} \cdot \left(-1\dfrac{5}{11}\right)$

14. $-5 \cdot \dfrac{17}{25}$

15. $2\dfrac{9}{10} \cdot 1\dfrac{1}{5}$

16. $\dfrac{6m}{13} \cdot \dfrac{2}{mn}$

17. $\dfrac{p}{3} \cdot \dfrac{1}{q}$

18. $\dfrac{2u}{v^2} \cdot \dfrac{3}{u}$

19. $\dfrac{4x}{3y} \cdot \dfrac{9y}{2x}$

20. $\dfrac{2a}{b} \cdot \dfrac{c}{2d}$

21. $\dfrac{rs}{9t} \cdot \dfrac{3}{s^2}$

22. $2x \cdot \dfrac{1}{4x^2}$

23. $\dfrac{x^2}{4y} \cdot \dfrac{16y^2}{3x}$

24. $\dfrac{2}{r} \cdot \dfrac{3}{r}$

Evaluate each expression if $a = -\dfrac{5}{6}$, $b = -3\dfrac{3}{8}$, and $c = \dfrac{7}{10}$. Write the product in simplest form.

25. bc

26. ac

27. $4\dfrac{2}{5}c$

28. $-2abc$

29. $-3\dfrac{3}{7}ab$

30. $2\dfrac{1}{9}abc$

31. **AIRPLANES** The fastest retired airliner, the Concorde, had the capability of cruising at speeds of up to 1450 mph. While cruising at this top speed, how far would the Concorde travel in $2\dfrac{1}{2}$ hours?

3-4 Skills Practice

Dividing Rational Numbers

Find the multiplicative inverse of each number.

1. $\dfrac{7}{12}$

2. $-\dfrac{3}{10}$

3. $\dfrac{1}{8}$

4. -64

5. $8\dfrac{1}{3}$

6. $-10\dfrac{2}{3}$

7. $-6\dfrac{5}{6}$

8. $1\dfrac{1}{8}$

Find each quotient. Write in simplest form.

9. $\dfrac{1}{3} \div \dfrac{7}{18}$

10. $-\dfrac{2}{5} \div \dfrac{4}{25}$

11. $-5 \div \dfrac{1}{7}$

12. $\dfrac{2}{3} \div \dfrac{2}{3}$

13. $\dfrac{4}{5} \div \left(-\dfrac{1}{15}\right)$

14. $\dfrac{19}{20} \div \dfrac{4}{5}$

15. $3 \div \dfrac{1}{4}$

16. $-15 \div \dfrac{1}{2}$

17. $\dfrac{4}{9} \div \dfrac{5}{12}$

18. $\dfrac{7}{10} \div \left(-\dfrac{4}{5}\right)$

19. $\dfrac{7}{12} \div \left(-1\dfrac{1}{6}\right)$

20. $1\dfrac{5}{8} \div \dfrac{5}{8}$

21. $12\dfrac{3}{5} \div 2\dfrac{7}{10}$

22. $-\dfrac{3}{11} \div \dfrac{6}{22}$

23. $\dfrac{1}{8} \div \dfrac{15}{16}$

24. $-12\dfrac{4}{5} \div \left(-1\dfrac{1}{15}\right)$

25. $1\dfrac{12}{13} \div \dfrac{25}{26}$

26. $-7\dfrac{1}{3} \div 2\dfrac{1}{5}$

27. $\dfrac{x}{6} \div \dfrac{x}{30}$

28. $\dfrac{12}{5x} \div \dfrac{6}{2x}$

29. $\dfrac{m}{16} \div \dfrac{mp}{7}$

30. $\dfrac{3r}{s} \div \dfrac{4rs}{s^2}$

31. $\dfrac{a}{b} \div \dfrac{5}{b}$

32. $\dfrac{2a}{b} \div \dfrac{3a^2}{b^2}$

33. $\dfrac{3}{5c} \div \dfrac{1}{10c}$

34. $\dfrac{pq}{6} \div \dfrac{q}{8}$

35. $\dfrac{x^2}{7} \div \dfrac{2x}{21}$

36. $\dfrac{gh}{6} \div \dfrac{36}{h}$

37. $\dfrac{3n}{2m} \div \dfrac{5n}{5m}$

38. $\dfrac{4b}{c} \div \dfrac{5bc}{c}$

3-4 Practice

Dividing Rational Numbers

Find each quotient. Write in simplest form.

1. $\frac{1}{2} \div \frac{1}{10}$

2. $-\frac{3}{8} \div \frac{9}{24}$

3. $-\frac{15}{16} \div \frac{7}{12}$

4. $\frac{17}{20} \div \left(-\frac{3}{10}\right)$

5. $-\frac{3}{8} \div \left(-\frac{3}{9}\right)$

6. $\frac{25}{32} \div \frac{15}{56}$

7. $0 \div \frac{17}{18}$

8. $-1\frac{1}{2} \div \frac{1}{4}$

9. $\frac{8}{9} \div \frac{22}{81}$

10. $8\frac{4}{9} \div 2\frac{1}{9}$

11. $4\frac{3}{5} \div \frac{2}{5}$

12. $-\frac{100}{63} \div \frac{10}{81}$

13. $18\frac{1}{3} \div \left(-4\frac{1}{6}\right)$

14. $-3\frac{2}{9} \div \frac{4}{27}$

15. $-2\frac{5}{6} \div \frac{3}{51}$

16. $4\frac{11}{12} \div 4\frac{5}{6}$

17. $\frac{2x}{3} \div \frac{1}{9}$

18. $\frac{a}{4} \div \frac{a}{8}$

19. $\frac{4k}{5} \div \frac{25}{2k}$

20. $\frac{ab}{8} \div \frac{b}{a}$

21. $\frac{2c}{b} \div \frac{4a}{b}$

22. $\frac{y}{x} \div y^2$

23. $\frac{3st}{r} \div \frac{4t}{r}$

24. $\frac{a^2}{b^2} \div \frac{c^2}{b^2}$

25. $-\frac{2x}{y} \div \frac{4}{y}$

26. $\frac{m^2}{2np} \div \frac{n}{4p}$

27. Evaluate $x \div y$ if $x = 3\frac{1}{2}$ and $y = \frac{3}{4}$.

28. Evaluate $w \div z$ if $w = \frac{6}{7}$ and $z = 3$.

29. **TRAVEL** What is the average speed that Robin must drive to reach her friend's house 170 miles away in $2\frac{1}{2}$ hours?

30. **SEWING** How many choir robes can be made from $20\frac{1}{4}$ yards of fabric if each robe needs $1\frac{1}{8}$ yards?

3-5 Skills Practice

Adding and Subtracting Like Fractions

Find each sum or difference. Write in simplest form.

1. $\dfrac{4}{15} + \dfrac{6}{15}$

2. $\dfrac{7}{12} + \dfrac{11}{12}$

3. $\dfrac{7}{10} + \dfrac{9}{10}$

4. $\dfrac{20}{21} - \dfrac{2}{21}$

5. $\dfrac{11}{12} - \dfrac{5}{12}$

6. $\dfrac{5}{8} + \dfrac{7}{8}$

7. $\dfrac{10}{11} + \dfrac{9}{11}$

8. $\dfrac{17}{30} - \dfrac{7}{30}$

9. $\dfrac{5}{6} + \dfrac{5}{6}$

10. $4\dfrac{4}{5} + 3\dfrac{2}{5}$

11. $20\dfrac{1}{25} + 1\dfrac{4}{25}$

12. $5\dfrac{11}{15} + 3\dfrac{14}{15}$

13. $26\dfrac{7}{12} + 11\dfrac{11}{12}$

14. $20\dfrac{3}{4} - 3\dfrac{1}{4}$

15. $25\dfrac{4}{5} - 3\dfrac{2}{5}$

16. $\dfrac{10}{15} - \dfrac{13}{15}$

17. $\dfrac{a}{6} + \dfrac{4a}{6}$

18. $\dfrac{7c}{16} + \dfrac{7c}{16}$

19. $\dfrac{25}{x} - \dfrac{17}{x}, x \neq 0$

20. $1\dfrac{1}{2y} - 2\dfrac{1}{2y}$

21. $\dfrac{7x}{9} - \dfrac{7x}{9}$

22. $\dfrac{3m}{5} + \dfrac{8m}{5}$

Evaluate each expression if $x = \dfrac{5}{8}$, $y = 1\dfrac{3}{8}$, and $z = \dfrac{1}{8}$.

23. $x + y$

24. $y - x$

25. $x - z$

26. $x + y + z$

3-5 Practice

Adding and Subtracting Like Fractions

Find each sum or difference. Write in simplest form.

1. $\dfrac{5}{7} + \dfrac{2}{7}$

2. $\dfrac{5}{11} - \dfrac{1}{11}$

3. $\dfrac{13}{20} - \dfrac{3}{20}$

4. $\dfrac{5}{16} + \dfrac{15}{16}$

5. $-\dfrac{19}{40} + \dfrac{21}{40}$

6. $-\dfrac{7}{9} - \dfrac{4}{9}$

7. $\dfrac{14}{23} - \dfrac{16}{23}$

8. $\dfrac{25}{36} + \left(-\dfrac{7}{36}\right)$

9. $\dfrac{21}{25} + \dfrac{9}{25}$

10. $10\dfrac{4}{7} + 11\dfrac{5}{7}$

11. $9\dfrac{3}{8} + 4\dfrac{1}{8}$

12. $-8\dfrac{7}{10} + 2\dfrac{3}{10}$

13. $23\dfrac{17}{20} - 4\dfrac{7}{20}$

14. $22\dfrac{3}{8} - 18\dfrac{5}{8}$

15. $7\dfrac{9}{10} + 3\dfrac{3}{10}$

16. $6\dfrac{1}{6} - 3\dfrac{5}{6}$

17. $5\dfrac{1}{4} + 3\dfrac{1}{4} + 9\dfrac{3}{4}$

18. $6\dfrac{7}{8} + \left(-7\dfrac{3}{8}\right)$

19. $\dfrac{h}{6} + \dfrac{4h}{6}$

20. $\dfrac{5c}{22} + \dfrac{5c}{22}$

21. $\dfrac{35}{d} - \dfrac{17}{d}, d \neq 0$

22. $\dfrac{4r}{9} + \dfrac{5r}{9}$

23. $\dfrac{6s}{t} + \dfrac{s}{t}$

24. $\dfrac{5p}{9} - \dfrac{4p}{9}$

25. $\dfrac{6r^2}{s^2} + \dfrac{5r^2}{s^2}$

26. $4\dfrac{5}{7a} - 2\dfrac{3}{7a}$

27. **PICTURE FRAMING** Matt plans to paste a picture that is $6\dfrac{7}{8}$ inches wide on a sheet of paper that is $8\dfrac{4}{8}$ inches wide. If he wants to have at least $\dfrac{5}{8}$ inch of margin on each side, will the picture fit? Explain.

3-6　Skills Practice

Adding and Subtracting Unlike Fractions

Find each sum or difference. Write in simplest form.

1. $\frac{4}{7} + \frac{1}{3}$

2. $\frac{2}{5} + \frac{3}{4}$

3. $\frac{1}{2} + \left(-\frac{3}{10}\right)$

4. $-\frac{5}{6} + \frac{7}{9}$

5. $\frac{5}{12} + \frac{23}{24}$

6. $\frac{10}{11} - \frac{1}{2}$

7. $\frac{4}{5} - \left(-\frac{1}{3}\right)$

8. $\frac{5}{6} - \frac{1}{12}$

9. $\frac{19}{20} + \frac{1}{4}$

10. $-\frac{9}{10} - \frac{1}{3}$

11. $\frac{13}{15} - \frac{2}{3}$

12. $\frac{7}{10} + \frac{1}{5}$

13. $-\frac{3}{8} + \frac{1}{6}$

14. $\frac{33}{100} - \frac{1}{10}$

15. $\frac{11}{12} - \left(-\frac{7}{8}\right)$

16. $\frac{4}{5} - \frac{1}{8}$

17. $5\frac{2}{3} + 2\frac{1}{6}$

18. $1\frac{7}{8} + 3\frac{1}{3}$

19. $3\frac{2}{3} - \frac{1}{9}$

20. $23\frac{3}{4} - 12\frac{5}{16}$

21. $-7\frac{1}{2} + \frac{3}{4}$

22. $2\frac{2}{3} + 1\frac{1}{4}$

23. $-12\frac{1}{2} - 17\frac{1}{2}$

24. $12\frac{1}{3} - \frac{3}{5}$

25. $11\frac{15}{16} - 7\frac{1}{2}$

26. $8\frac{5}{9} + 1\frac{1}{6}$

27. $-7\frac{1}{2} + 3\frac{1}{7}$

28. $60\frac{1}{2} + \left(-37\frac{1}{6}\right)$

29. $8\frac{2}{3} - 3\frac{1}{3}$

30. $-21\frac{7}{16} + 13\frac{1}{4}$

3-6 Practice

Adding and Subtracting Unlike Fractions

Find each sum or difference. Write in simplest form.

1. $\frac{9}{10} + \frac{1}{2}$

2. $\frac{7}{8} + \frac{1}{10}$

3. $-\frac{3}{4} + \frac{5}{16}$

4. $\frac{4}{5} - \frac{2}{6}$

5. $\frac{5}{8} - \frac{3}{16}$

6. $\frac{1}{3} + \frac{5}{36}$

7. $\frac{7}{10} - \frac{14}{100}$

8. $\frac{17}{21} - \frac{4}{6}$

9. $\frac{11}{14} - \frac{1}{6}$

10. $\frac{4}{15} - \left(-\frac{3}{12}\right)$

11. $\frac{7}{15} + \frac{3}{6}$

12. $-\frac{7}{8} + \frac{9}{10}$

13. $10\frac{1}{2} + 7\frac{1}{3}$

14. $7\frac{1}{2} - 2\frac{7}{10}$

15. $8\frac{1}{6} + 5\frac{3}{4}$

16. $7\frac{7}{12} - 5\frac{1}{3}$

17. $6\frac{4}{5} + \left(-2\frac{3}{8}\right)$

18. $16\frac{3}{5} + 3\frac{11}{15}$

19. $18\frac{3}{5} - 7\frac{1}{4}$

20. $12\frac{2}{7} - 3\frac{5}{6}$

21. $2\frac{5}{8} + 6\frac{3}{4}$

22. $29\frac{8}{33} + \left(-3\frac{1}{3}\right)$

23. $-6\frac{2}{7} - 5\frac{3}{14}$

24. $-16\frac{2}{7} - 3\frac{20}{21}$

25. $-10\frac{1}{9} + 9\frac{7}{45}$

26. $\frac{1}{3} + \frac{5}{6} + \frac{1}{2}$

27. $9\frac{2}{7} - 11\frac{18}{21}$

28. $-17\frac{2}{3} - \left(-5\frac{4}{18}\right)$

29. $11\frac{3}{16} - 5\frac{1}{12}$

30. $\frac{64}{143} - \frac{21}{208}$

31. **SEWING** The inseam on Juan's pants is $34\frac{1}{4}$ inches. If he has them shortened by $2\frac{7}{8}$ inches, what is the new length?

4-2 Skills Practice

Simplifying Algebraic Expressions

Identify the terms, like terms, coefficients, and constants in each expression.

1. $7a + a$

2. $3k + g - k$

3. $m + 3m + 8$

4. $10b - bc + 1 + 3bc$

5. $9j + 8j - 4 - 7j$

6. $6y + 3x + 6y - 2x$

7. $3q + 2 - 7p$

8. $18 + 7x - 12 + 5x$

9. $12a + 3b + 18 - 9a$

Simplify each expression.

10. $13c - 7 + c - d$

11. $5h + h - 4h + 1 - 2h$

12. $2(v - 5) + 7v + 4$

13. $5(r + 9) - 5$

14. $1 - 4(u - 1)$

15. $-7(w - 4) + 3w - 27$

16. $-8 - 7(y + 2)$

17. $-18(c - 1) - 18$

18. $12(n - 4) - 3n$

19. $5m - 9 + 4m$

20. $-7 + g + 1 - 6g$

21. $x - 9x + 3 + 8x - 3$

22. $6(r - 4) + r + 30 - 7r$

23. $-5 + 5a - 4 - 2a + 3a$

24. $21 - 8(v + 3) + 3 + 7v$

25. $4x - 9 + 3x + 6 - 9x - 4$

26. $p - 2 + 1 - p + 1 + 2p$

27. $-11f + 6 - f + 4 + 13f - 9$

28. $3(d - 4) + 2 - 2d + 1 - d$

29. $1 - s + 2 + 2s - 3s + 1$

30. $5 - 9k + 1 + k - 2(7 - k)$

4-2 Practice

Simplifying Algebraic Expressions

Identify the terms, like terms, coefficients, and constants in each expression.

1. $6y - 4 + y$

2. $8u + 2u - 3u$

3. $-12h + 5g + 8 - g$

4. $-21w + 5 + 3w - 1$

5. $8a + b - 3a + 4b$

6. $f - 3fg + 2g - fg + 1$

Simplify each expression.

7. $-8q + 6 + 5q - 3$

8. $h + 5h - 3 - 6h$

9. $2a - 5(a + 1)$

10. $b - 2(b - 2)$

11. $9 - t - 3(t + 3)$

12. $-8 + 5(g + 2) - 2$

13. $12m + 9 - 2m - 16$

14. $4(y - 3) + 9 - 3y$

15. $r + r + r + r + r$

16. $-11x + 4 + 8x - 4 + 3x$

17. $-14y + 12(x + y) - 12x$

18. $19g - 4h + 4 - 20(g - 1)$

19. $-5(c + d) - 4d + 5c - d$

20. $(8 - b)(-3) + 6b + 12 - 10b$

21. $-p + q + 2(p + q) - p - q$

22. $-55n + 28n + 21n + 7n - n$

23. $-12z + 4(z - 9) + 30 + z$

Write an expression in simplest form that represents the total amount in each situation.

24. LUNCH You bought 3 pieces of chicken that cost x dollars each, a salad for $3, and a drink for $1.

25. SOCCER Sal has scored g goals this season. Ben has scored four times as many goals as Sal. Chun has scored three fewer goals than Ben.

4-3 Skills Practice

Solving Equations by Adding or Subtracting

Solve each equation. Check your solution.

1. $r + 1 = -5$ **2.** $h + 8 = 6$ **3.** $t - 3 = -11$ **4.** $p - 5 = 9$

5. $w + 9 = -9$ **6.** $x - 9 = -9$ **7.** $a + 7 = -7$ **8.** $m + 9 = -7$

9. $q - 4 = 5$ **10.** $b + 2 = 3$ **11.** $n - 11 = 1$ **12.** $r - 1 = -3$

13. $c + 6 = 1$ **14.** $v - 3 = -7$ **15.** $z + 3 = 0$ **16.** $s - 8 = -1$

17. $y - 7 = -5$ **18.** $u - 10 = -2$ **19.** $g + 1 = 10$ **20.** $k + 4 = -9$

21. $w + 12 = -4$ **22.** $z - 8 = -8$ **23.** $d - 11 = 1$ **24.** $h + 3 = 10$

25. $r + 10 = -6$ **26.** $y + 1 = 4$ **27.** $f - 6 = 6$ **28.** $d - 2 = -8$

29. $j + 11 = 4$ **30.** $m - 10 = 4$ **31.** $q + 3 = -5$ **32.** $g - 4 = 0$

33. $a - 12 = -19$ **34.** $c + 5 = 2$ **35.** $h - 9 = 12$ **36.** $p + 14 = -1$

37. $v + 13 = -11$ **38.** $x + 8 = -1$ **39.** $y + 12 = -10$ **40.** $k - 16 = 7$

41. $d - 15 = -14$ **42.** $g - 12 = 10$ **43.** $b + 13 = -20$ **44.** $f - 15 = -1$

45. $q + 8 = 13$ **46.** $w - 4 = -15$ **47.** $r + 10 = -13$ **48.** $t - 11 = 11$

49. $j - 9 = -8$ **50.** $k + 2 = -15$ **51.** $n + 12 = 0$ **52.** $y + 9 = 14$

4-3   Practice

Solving Equations by Adding or Subtracting

Solve each equation. Check your solution.

1. $z + 6 = -5$ **2.** $x - 8 = -3$ **3.** $c - 2 = 21$ **4.** $v + 9 = 0$

5. $q + 10 = -30$ **6.** $w + 15 = 0$ **7.** $z + 12 = -19$ **8.** $b - 11 = 8$

9. $a - 12 = 0$ **10.** $r + 11 = 12$ **11.** $p + (-9) = 33$ **12.** $n - 16 = -16$

13. $s + 13 = -5$ **14.** $t - (-15) = 21$ **15.** $r - 14 = -23$ **16.** $m + (-3) = 9$

17. $d - 19 = 1$ **18.** $y + 30 = -1$ **19.** $u - 21 = 0$ **20.** $k - 18 = 2$

21. $f - 23 = 23$ **22.** $g - 24 = -24$ **23.** $h + 35 = 7$ **24.** $j + 40 = 25$

25. $x + 3 = -15$ **26.** $c + 22 = -27$ **27.** $v - 18 = -4$ **28.** $b - 41 = -30$

29. $h - 10 = 19$ **30.** $y - (-12) = 0$ **31.** $g + 58 = 9$ **32.** $n + 29 = 4$

33. $j + (-14) = 1$ **34.** $p - 21 = -2$ **35.** $k - (-13) = -8$ **36.** $m + 33 = 16$

37. SAVINGS ACCOUNT Jhumpa has $55 in her savings account. This is $21 more than David. Write and solve an equation to find the amount David has in his savings account.

38. WEATHER The temperature fell 16° between noon and 3:00 P.M. At 3:00, the temperature was −3°F. Write and solve an equation to determine the temperature at noon.

4-4 Skills Practice

Solving Equations by Multiplying or Dividing

Solve each equation. Check your solution.

1. $3x = 24$

2. $\dfrac{m}{-5} = -15$

3. $-4f = 16$

4. $\dfrac{u}{2} = 12$

5. $-6a = 6$

6. $\dfrac{s}{-1} = 10$

7. $-2y = -2$

8. $-7z = 7$

9. $\dfrac{n}{8} = -24$

10. $-4r = -12$

11. $-9h = 81$

12. $\dfrac{c}{-10} = 1$

13. $\dfrac{v}{-15} = -15$

14. $\dfrac{m}{12} = 0$

15. $-12g = 12$

16. $\dfrac{w}{-4} = 0$

17. $-1f = 11$

18. $\dfrac{r}{-1} = 22$

19. $8d = -16$

20. $\dfrac{r}{15} = 45$

21. $25k = -200$

22. $-3p = 18$

23. $7j = -63$

24. $\dfrac{y}{-10} = 10$

25. $\dfrac{x}{-8} = -1$

26. $5g = -20$

27. $\dfrac{p}{6} = 0$

28. $7y = 7$

29. $-6q = -30$

30. $-12c = -60$

31. $-9b = 90$

32. $-4k = -120$

33. $2r = 0$

34. $-1t = 19$

35. $\dfrac{n}{-12} = 12$

36. $-15j = 120$

37. $\dfrac{u}{-11} = 11$

38. $5c = 85$

39. $-9q = -36$

40. $9z = -144$

4-4 Practice

Solving Equations by Multiplying or Dividing

Solve each equation. Check your solution.

1. $8y = 56$

2. $\dfrac{w}{4} = 12$

3. $-3u = -12$

4. $\dfrac{r}{-5} = 15$

5. $9d = -9$

6. $-8f = 0$

7. $\dfrac{n}{-1} = 31$

8. $\dfrac{v}{14} = -7$

9. $-1b = 24$

10. $-12h = -72$

11. $\dfrac{r}{24} = -5$

12. $\dfrac{p}{-6} = -3$

13. $-15x = 90$

14. $-4g = -20$

15. $\dfrac{z}{20} = -1$

16. $11t = 0$

17. $23g = -92$

18. $-7d = -28$

19. $\dfrac{m}{-15} = 7$

20. $9k = -9$

21. $6w = 0$

22. $-4r = 120$

23. $\dfrac{u}{12} = 1$

24. $-11q = -99$

25. $16y = -192$

26. $\dfrac{n}{-8} = 0$

27. $-7j = 84$

28. $-21p = -231$

Write and solve an equation for each sentence.

29. The product of a number and -6 is -54.

30. The quotient of a number and 6 is -14.

31. **CLASS REPORTS** Each student needs 12 minutes to give a report. A class period is 48 minutes long. Write and solve an equation to determine the number of students who could give a report in one class period.

32. **COOKING** One pound of ground beef makes four hamburger patties. Write and solve an equation to determine how many pounds of beef are needed to make 36 hamburgers.

4-5 Skills Practice

Solving Two-Step Equations

Solve each equation. Check your solution.

1. $3x + 10 = 1$

2. $\frac{a}{5} + 8 = 9$

3. $8w - 12 = -4$

4. $\frac{r}{2} + 6 = 5$

5. $18 - 2q = 4$

6. $3j - 20 = 16$

7. $\frac{u}{12} - 8 = -8$

8. $7p + 11 = -31$

9. $12d + 15 = 3$

10. $4c + 20 = 0$

11. $\frac{n}{2} - 9 = -5$

12. $10b - 19 = 11$

13. $2h + 10 = -12$

14. $6k - 9 = 15$

15. $\frac{w}{-5} - 4 = -2$

16. $12 - 7y = -2$

17. $11 - 3g = 32$

18. $12s + 13 = 25$

19. $2z - 4 - z = 4$

20. $10 - 5h + 2 = 32$

21. $\frac{r}{-7} - 5 = -6$

22. $-4a + 5 - 2a - 9 = 44$

23. $\frac{w}{-3} + 6 - 1 = 2$

24. $7k - 8k = 1$

25. $7f - 24 = 25$

26. $6 - \frac{m}{6} - 8 = 0$

27. $10 - d = 19$

28. $9x + 5 - 4x = -20$

29. $3 - 4t + 11 = 2$

30. $\frac{a}{3} - 4 + 9 = 7$

31. $6q - 4 = -16$

32. $\frac{m}{8} - 12 - 3 = -12$

33. $5b + 6 - 6b + 2 = 19$

4-5 Practice

Solving Two-Step Equations

Solve each equation. Check your solution.

1. $6p + 22 = 10$

2. $\dfrac{r}{3} - 4 = 2$

3. $5d - 9 = -24$

4. $21q - 11 = 10$

5. $\dfrac{v}{-6} + 1 = 0$

6. $7h + 20 = -8$

7. $8k - 40 = 16$

8. $\dfrac{w}{2} - 16 = 5$

9. $\dfrac{s}{4} - 5 = 1$

10. $\dfrac{x}{8} + 7 = 9$

11. $\dfrac{z}{10} - 20 = -20$

12. $\dfrac{r}{-2} + 11 = 15$

13. $9q + 10 = 118$

14. $\dfrac{n}{5} - 4 = -10$

15. $6w - 125 = 1$

16. $\dfrac{r}{3} - 16 = 2$

17. $9y - 11 - 5y = 25$

18. $20 - 15d = 35$

19. $\dfrac{u}{-9} - 8 = -4$

20. $-6h + 4 - 3 + h = 11$

21. $5p - 4p = 7$

22. $18 - \dfrac{x}{3} = -7$

23. $21 + 9j - 10 = -277$

24. $12b - 9 + 2b - b = -87$

25. $1 + \dfrac{a}{-9} - 4 = 0$

26. $4w - w - 26 = 19$

27. $5 - 4y + y - 1 = -23$

28. **RENTAL AGREEMENTS** A furniture rental store charges a down-payment of $100 and $75 per month for a table. Hilde paid $550 to rent the table. Solve $75n + 100 = 550$ to find the number of months Hilde rented the table.

29. **BUSINESS** At work, Jack must stuff 1000 envelopes with advertisements. He can stuff 12 envelopes in one minute, and he has 112 envelopes already finished. Solve $1000 = 12n + 112$ to find how many minutes it will take Jack to complete the task.

4-6 Skills Practice

Writing Equations

Translate each sentence into an equation. Then find each number.

1. Eleven less than 5 times a number is 24.

2. The quotient of a number and −9 increased by 10 is 11.

3. Five less than the product of −3 and a number is −2.

4. Fifteen more than twice a number is −23.

5. The difference between 5 times a number and 4 is 16.

6. Nine more than −8 times a number is −7.

7. The difference between 12 and ten times a number is −28.

8. Seven more than three times a number is 52.

9. Eleven less than five times a number is 19.

10. Thirteen more than four times a number is −91.

11. Seven less than twice a number is 43.

Solve each problem by writing and solving an equation.

12. **SHOPPING** The total cost of a suit and 4 ties is $292. The suit cost $200. Each tie cost the same amount. Find the cost of one tie.

13. **AGES** Mary's sister is 7 years older than Mary. Their combined ages add up to 35. How old is Mary?

4-6 Practice

Writing Equations

Translate each sentence into an equation. Then find each number.

1. Eight less than 7 times a number is −29.

2. Twenty more than twice a number is 52.

3. The difference between three times a number and 11 is 10.

4. One more than the difference between 18 and seven times a number is −9.

5. Eight times a number plus 6 less than twice the number is 34.

6. 26 more than the product of a number and 17 is −42.

7. Twelve less than the quotient of a number and 8 is −1.

Solve each problem by writing and solving an equation.

8. **ANIMAL TRAINING** Last summer, Gary trained 32 more dogs than Zina. Together they trained 126 dogs. How many dogs did Gary train?

9. **SALES** Julius sold five times as many computers as Sam sold last year. In total, they sold 78 computers. How many computers did Julius sell?

10. **TRACK** In one season, Ana ran 18 races. This was four fewer races than twice the number of races Kelly ran. How many races did Kelly run?

11. **BASEBALL** André hit four more home runs than twice the number of home runs Larry hit. Together they hit 10 home runs. How many home runs did André hit?

12. **FUNDRAISING** The sixth grade has collected $116 for a local animal shelter. Their goal is to collect $500. They have 3 weeks left. How much money must they collect each week?

5-1 Skills Practice

Perimeter and Area

Find the perimeter and area for each figure.

1.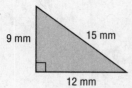

9 mm   15 mm   12 mm

2.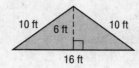

10 ft   6 ft   10 ft   16 ft

3.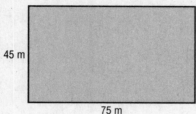

45 m   75 m

4.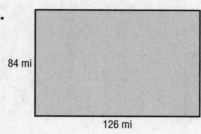

84 mi   126 mi

5. a right triangle with legs of 7 inches and 24 inches and a hypotenuse of 25 inches

6. a rectangle that is 21 inches long and 13 inches wide

7. a square that is 25 centimeters on each side

Find the missing dimension for each figure.

8.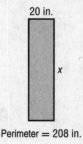

20 in. x Perimeter = 208 in.

9.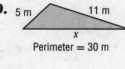

5 m   11 m x Perimeter = 30 m

10.

12 mi   20 mi x Area = 96 mi^2

11.

8 ft x Area = 216 ft^2

12. The perimeter of a rectangle is 100 centimeters. Its width is 9 centimeters. Find its length.

13. The area of a rectangle is 319 square kilometers. Its width is 11 kilometers. Find its length.

14. The perimeter of a triangle is $37\frac{1}{2}$ yards. Two of the sides each measure $12\frac{1}{8}$ yards. Find the third side.

5-1   Practice

Perimeter and Area

Find the perimeter and area for each figure.

1.

14 cm
50 cm
48 cm

2.

60 yd
36 yd   75 yd
45 yd

3.

61 mi
54 mi

4.

48 mm
48 mm

5. a right triangle with legs of 24 feet and 41.57 feet and a hypotenuse of 48 feet

6. a rectangle that is 92 meters long and 18 meters wide

7. a rectangle that is 30 inches long and 29 inches wide

Find the missing dimension for each figure.

8.

53 mi   46 mi
x
Perimeter = 160 mi

9.

36 cm   60 cm
x
Area = 864 cm^2

10.

x
25 ft
Area = 1125 ft^2

11.

13 m
x
Perimeter = 68 m

12. **GEOMETRY** The area of a rectangle is 1260 square inches. Its length is 36 inches. Find the width.

13. **GEOMETRY** The area of a triangle is 672 square yards. Its base is 42 yards. Find the height.

5-3 Skills Practice

Inequalities

Write an inequality for each sentence.

1. More than 100,000 fans attended the opening football game at The Ohio State University.

2. Her earnings at $16 per hour were no more than $96.

3. A savings account decreased by $50 is now less than $740.

4. A number increased by 7 is at least 45.

For the given value, state whether each inequality is *true* or *false*.

5. $\frac{18}{c} < 9$, $c = 2$ 6. $\frac{x}{5} \geq 3$, $x = 5$

7. $6k \geq 42$, $k = 7$ 8. $10 - x < 3$, $x = 7$

9. $11 + n < 32$, $n = 4$ 10. $9 + c > 19$, $c = 10$

Graph each inequality on a number line.

11. $a < 6$

12. $t \geq -2$

13. $d \leq 3$

14. $b \geq 10$

15. $x \geq -7$

16. $x > 2$

Write the inequality for each graph.

17.

18.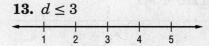

19.

20.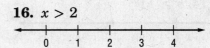

21.

22.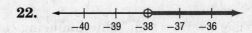

5-3 Practice

Inequalities

Write an inequality for each sentence.

1. More than 3400 people attended the flea market.

2. Her earnings at $11 per hour were no more than $121.

3. The 10-km race time of 84 minutes was at least twice as long as the winner's time.

4. A savings account increased by $70 is now more than $400.

For the given value, state whether each inequality is *true* or *false*.

5. $9 - x > 3, x = 6.5$ **6.** $9.5 + n < 19, n = 10$

7. $3k < 27\frac{1}{2}, k = 8$ **8.** $21 \le 4c, c = 5.2$

9. $\frac{x}{4} \le 8, x = 32$ **10.** $\frac{9}{c} > 2, c = 3\frac{1}{2}$

Graph each inequality on a number line.

11. $a < -2$

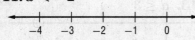

12. $t > -6$

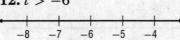

13. $d \ge 7$

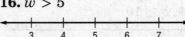

14. $b \ge 11$

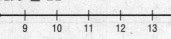

15. $x \le -8$

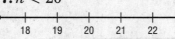

16. $w > 5$

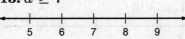

17. $n < 20$

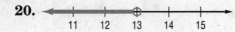

18. $b \le -4$

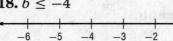

19. $a \ge -6$

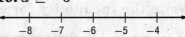

Write the inequality for each graph.

20.

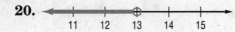

21.

22.

23.

24. HIPPOS The average time a human being can hold their breath underwater is 1 minute. A hippo can hold its breath underwater for at least 5 times as long as a human. Write an inequality that represents how long a hippo can hold its breath underwater.

25. CHARITY In the first hour of a charity auction, $4800 was raised. This was at most $1200 more than was raised in the second hour of the auction. Write an inequality that represents the amount raised in the second hour.

5-4 Skills Practice

Solving Inequalities

Solve each inequality. Check your solution.

1. $p + 9 > 13$ **2.** $t + 7 < -4$ **3.** $-12 \geq 7 + x$ **4.** $f + (-7) \leq 9$

5. $5 > -3 + y$ **6.** $r + 7 \leq -3$ **7.** $b - 15 > 11$ **8.** $z + (-4) < -8$

9. $j - 4 \leq -10$ **10.** $-5 > h - 3$ **11.** $13 > w - (-14)$ **12.** $g - 7 > -4$

13. $-15 \leq d + (-2)$ **14.** $2 + c \leq -8$ **15.** $15 > c + 3$ **16.** $j + 9 \leq -10$

Solve each inequality. Then graph the solution on a number line.

17. $-8x > 16$

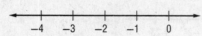

18. $7y < -35$

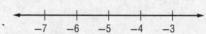

19. $12a \geq -24$

20. $-12 \leq 4a$

21. $-6z < -18$

22. $14 > -2k$

23. $5 > \dfrac{x}{-2}$

24. $\dfrac{r}{-3} \leq -4$

25. $-10t \geq 200$

26. $\dfrac{y}{7} < 2$

27. $\dfrac{-1}{2} x \leq -6$

28. $\dfrac{b}{-3} \leq 6$

29. SHOPPING Chantal would like to buy a new pair of running shoes. Shoes that she likes start at $85. If she has already saved $62, what is the least amount she must still save?

5-4 Practice

Solving Inequalities

Solve each inequality. Check your solution.

1. $-6 \geq g + 4$

2. $15 + d > 10$

3. $p + (-8) \leq -12$

4. $-13 < k - (-16)$

5. $-1 + s \leq 5$

6. $12 > w - (-0.3)$

7. $-1\frac{7}{8} < d + (-2)$

8. $z - 0.9 > -4.8$

9. $b - \frac{1}{5} < 3\frac{1}{10}$

Solve each inequality. Then graph the solution on a number line.

10. $24 \geq \dfrac{g}{-4}$

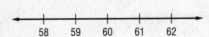

-98 -97 -96 -95 -94

11. $-78 > 6h$

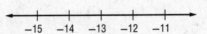

-15 -14 -13 -12 -11

12. $\dfrac{f}{-5} < -12$

58 59 60 61 62

13. $100 \geq -4s$

-27 -26 -25 -24 -23

14. $\dfrac{p}{-36} < 6$

-218 -217 -216 -215 -214

15. $-4 > \dfrac{c}{-3.5}$

12 13 14 15 16

16. $-24 < \dfrac{1}{2}b$

-50 -49 -48 -47 -46

17. $-3 \leq \dfrac{c}{-1.5}$

2 3 4 5 6

18. TRANSPORTATION A certain minivan has a maximum carrying capacity of 1200 pounds. If the luggage weighs 150 pounds, what is the maximum weight allowable for passengers?

19. DISCOUNTS To qualify for a store discount, Jorge's soccer team must spend at least $560 for new jerseys. The team needs 20 jerseys.

    a. Write an inequality to represent how much the team should spend on each jersey to qualify for the discount.

    b. How much should the team spend for each jersey?

5-5 Skills Practice

Solving Multi-Step Equations and Inequalities

Solve each equation. Check your solution.

1. $2(g - 7) = 16$

2. $5(x + 2) = 30$

3. $3(2d + 7) = 39$

4. $4(a - 2) = 3(a + 4)$

5. $3(f + 2) + 9 = 13 + 5f$

6. $2(x - 4) = 3(1 + x)$

7. $2n + 5 = 4(n + 2) - n$

8. $4(x + 3) = x$

9. $2(c - 3) = 76$

10. $7(x - 2) = 5(x + 2)$

11. $2(6x + 1) = 4(x - 5) - 2$

12. $4(2b - 6) + 11 = 8b - 13$

13. $6 + 6(2t - 1) = 3 + 12t$

14. $9t - 21 = 3(t - 7) + 6t$

15. $3(4k + 14) = 10k - 2(k - 7)$

Solve each inequality. Graph the solution on a number line.

16. $3x + 9 < 18$

17. $5 + 2c < -9$

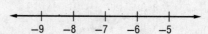

18. $4x - 3 < 2 - x$

19. $3(n + 2) < 24$

20. $11 + 2b \leq 3(2 - b)$

21. $\frac{m}{3} + 5 \geq 2$

22. $\frac{1}{2}(8 - x) > 6$

23. $\frac{c}{4} + 7 \geq 5$

24. $y - 3 < 5y + 1$

25. $20 - 2n > 26$

26. $\frac{1}{3}(x - 6) < 2$

27. $5 - 2k \leq 15$

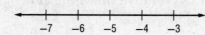

28. $-2(3 + t) < -8$

29. $\frac{n}{4} - 9 > 5$

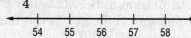

5-5 Practice

Solving Multi-Step Equations and Inequalities

Solve each equation. Check your solution.

1. $4(j - 7) = 12$

2. $5(2k + 10) = 40$

3. $7(2p + 3) - 8 = 6p + 29$

4. $7(g - 4) = 3$

5. $3(4c + 5) = 24$

6. $2(a - 1) = 3(a + 1)$

7. $3(x - 3) = 5(1.5 + x)$

8. $2(1.5m + 3) = 3.5m - 1$

9. $a - \frac{5}{10} = 2a - \frac{3}{5}$

10. $2.2x - 5 = 2(1.4x + 3)$

11. $\frac{d}{0.2} = 3d + 2.1$

12. $5n + 3 = 2(n + 2) - 3n$

13. $\frac{2}{3}a + 2 = \frac{1}{3}(4a + 1)$

14. $y - 7 = \frac{1}{4}(y + 2)$

Solve each inequality. Graph the solution on a number line.

15. $\frac{2}{3}(12 - x) > 4$

16. $\frac{1}{2}(8 - c) < 7.5$

17. $\frac{c}{3} + 7 > 5\frac{1}{2}$

18. $7 + 2p < -14$

19. $-3(x + 3) > 7.5$

20. $5 - 3c \le c + 17$

21. $2(n - 5) \le -7$

22. $\frac{18 - n}{2} \le 6$

23. GEOMETRY The perimeter of a rectangle is 80 feet. Find the dimensions if the length is 5 feet longer than four times the width. Then find the area of the rectangle.

24. NUMBER THEORY Five times the sum of three consecutive integers is 150. What are the integers?

25. STATE FAIR Admission to the state fair costs $5 and each ride costs $0.75. If Ahmed wants to spend no more than $14 at the fair, how many rides can he ride?

6-1 Skills Practice

Ratios

Express each ratio as a fraction in simplest form.

1. 8 pencils to 12 pens

2. 42 textbooks to 28 students

3. 27 rooms to 48 windows

4. 15 angel fish to 75 fish

5. 75 cats to 100 dogs

6. 6 aces out of 24 serves

7. 42 flowers to 7 vases

8. 14 boys to 21 girls

9. 50 nickels out of 125 coins

10. 9 children to 24 adults

11. 3 gallons to 15 quarts

12. 30 feet to 11 yards

13. 18 inches to 3 feet

14. 1 yard to 1 foot

15. 2 cups to 4 pints

16. 12 seconds to 1 minute

17. 3 pounds to 15 ounces

18. 15 inches to 2 yards

19. 1 pint to 4 quarts

20. 3 minutes to 1 hour

21. 8 ounces to 2 pounds

22. 7 quarts to 2 gallons

23. 6 ounces to 1 cup

24. 2 feet to 3 inches

6-1　Practice

Ratios

Express each ratio as a fraction in simplest form.

1. 56 pencils to 64 erasers

2. 25 calculators to 20 students

3. 36 cassettes to 60 CDs

4. 18 minnows to 27 fish

5. 26 tents to 65 campers

6. 49 apples out of 63 fruit

7. 45 out of 75 days

8. 60 forks to 144 spoons

9. 112 out of 200 pages

10. 36 balls to 81 players

11. 6 pounds to 256 ounces

12. 5 hours to 720 minutes

13. 9 gallons to 48 quarts

14. 24 feet to 30 yards

15. 420 seconds to 10 minutes

16. 96 inches to 9 feet

17. 64 cups to 50 pints

18. 35 pints to 7 gallons

19. 4 inches to 3 yards

20. 780 seconds to 1 hour

21. HOMECOMING At a homecoming game, there are 630 students and 1,080 alumni in attendance. Express the ratio of students to alumni as a fraction in simplest form. Explain its meaning.

6-2 Skills Practice

Unit Rates

Express each rate as a unit rate. Round to the nearest tenth or nearest cent, if necessary.

1. $9 for 6 cans of soup

2. $39 for a case of 75 bananas

3. 108 miles in 6 days

4. 51 meters in 8 seconds

5. 21 new pairs of sneakers in 7 years

6. 52 feet for 8 costumes

7. 40 sneezes in 20 minutes

8. $2702 from 28 people

9. **JUICE** A 64-ounce container of sports juice costs $6.50. A 48-ounce container of the same juice costs $4.25. Which size container is the better buy?

10. **KNITTING** Charmaine can knit 15 rows in 22 minutes. How many full rows could she knit in 90 minutes?

11. **STUDENTS** There are 156 sixth graders and 7 sixth-grade teachers. There are 120 fifth graders and 5 fifth-grade teachers. Which grade has the lower student to teacher ratio?

12. **PHONES** Cell phone Company X charges $15 for 120 minutes. Cell phone Company Y charges $25.95 for 300 minutes. Which company has the better per minute rate?

13. **ANIMALS** During normal sleep, a bear's heart beats about 50 times a minute. In its deepest state of hibernation, a bear's heart may beat 50 times in 6 minutes. During deep hibernation, how many times would the bear's heart beat in 45 minutes?

14. **PLANES** An airplane traveled 1536 miles in 3 hours. At this same rate, how far could the plane travel in 8 hours?

15. **ICE CREAM** An ice cream store makes 144 quarts of ice cream in 8 hours. How many quarts could be made in 12 hours?

6-2 Practice

Unit Rates

Express each rate as a unit rate. Round to the nearest tenth or nearest cent, if necessary.

1. $4.60 for 5 cans of soup

2. $51 for a box of 75 tiles

3. 652 miles in 9 days

4. 116 meters in 12 seconds

5. 176 new employees in 22 years

6. 34 yards for 6 costumes

7. 55 pages in 25 minutes

8. $3015 from 36 people

9. **CAMP** Happy Times Summer Camp has 356 campers and 38 counselors. PlayDay Summer Camp has 219 campers and 28 counselors. Which camp has the lower rate of campers to counselors?

10. **ROLLER COASTER** A roller coaster can accommodate 346 riders in 20 minutes. How many riders could ride in 90 minutes?

11. **BAGELS** The bakers at Joey's Bagels can make 340 bagels in 4 hours. How many bagels could the bakers make in 10 hours?

12. **CEREAL** The prices for various sizes of Health Crunch cereal are given in the table at the right. Which size has the best cost per ounce?

Size (oz)	Price
11	$4.75
15	$4.85
19.1	$5.89

13. **MUSIC** The Music Factory offers 45-minute music lessons for $40. The Music Makers offers 60-minute lessons for $55. Which is the better deal?

14. **RUNNING** Leslie ran a 5-kilometer race in 22 minutes. Jorge ran a 2-kilometer race in 8.5 minutes. Which runner ran at the faster rate?

15. **SEWING** It took Michala 4 hours to sew 9 scarves. How many scarves could she make in 24 hours?

6-3   Skills Practice

Converting Rates and Measurements

Convert each rate using dimensional analysis. Round to the nearest hundredth if necessary.

1. 12 m/min = ■ cm/s

2. 8 qt/min = ■ gal/h

3. 44 yd/s = ■ mi/h

4. 10 c/min = ■ qt/h

5. 32 ft/h = ■ yd/day

6. 56 mi/h = ■ ft/min

7. 40 cm/s = ■ m/min

8. 180 in./min = ■ yd/h

9. 220 mi/h = ■ yd/min

10. 3 km/h = ■ m/s

Complete each conversion. Round to the nearest hundredth if necessary.

11. 5 m ≈ ■ yd

12. 1500 mi ≈ ■ m

13. 20 L ≈ ■ gal

14. 70 kg ≈ ■ lb

15. 42 in. ≈ ■ cm

16. 38 ft ≈ ■ m

17. 2 kg ≈ ■ oz

18. 55 oz ≈ ■ g

19. 1200 km ≈ ■ ft

20. 18 qt ≈ ■ ml

Convert each rate using dimensional analysis. Round to the nearest hundredth if necessary.

21. 10 km/h ≈ ■ mi/h

22. 40 gal/s ≈ ■ L/h

23. 24 yd/min ≈ ■ cm/min

24. 16 L/h ≈ ■ gal/h

25. 300 mi/h ≈ ■ km/min

26. 120 mi/day ≈ ■ km/week

27. 100 L/day ≈ ■ qt/h

28. 55 m/min ≈ ■ in./s

29. 12 pt/h ≈ ■ L/min

30. 2800 mi/h ≈ ■ km/h

6-3 Practice

Converting Rates and Measurements

Convert each rate using dimensional analysis. Round to the nearest hundredth if necessary.

1. 18 m/min = ■ cm/s

2. 5.7 gal/h = ■ c/min

3. 264 yd/s = ■ mi/h

4. 2 qt/min = ■ gal/h

5. 99 in./s = ■ mi/day (1 day = 24 h)

6. 154 mi/h = ■ in./s

7. 44 mi/m = ■ ft/s

8. 15 oz/min = ■ gal/h

Complete each conversion. Round to the nearest hundredth if necessary.

9. 10 cm ≈ ■ in.

10. 300 gal ≈ ■ L

11. 250 g ≈ ■ oz

12. 5.5 kg ≈ ■ lb

13. 145 m ≈ ■ mi

14. 9.5 L ≈ ■ pt

15. 13 yd ≈ ■ m

16. 1,095 mi ≈ ■ km

Convert each rate using dimensional analysis. Round to the nearest hundredth if necessary.

17. 88 mi/h ≈ ■ km/min

18. 10 ft/min ≈ ■ m/h

19. 165 L/h ≈ ■ qt/min

20. 26 yd/s ≈ ■ km/h

21. 474 gal/day ≈ ■ L/week

22. 33.6 m/s ≈ ■ ft/min

23. 22 fl oz/min ≈ ■ mL/s

24. 299 km/h ≈ ■ mi/min

25. **TRACK AND FIELD** Rita sprinted 77 feet in 10 seconds. How many miles per hour is this?

26. **TRAVEL** Lisa is traveling to Europe. The information from the airlines said that she is only allowed to check 25 kilograms worth of baggage. How many pounds is this?

27. **SPACE SHUTTLE** The space shuttle travels at an orbital speed of about 17,240 miles per hour. How many meters per minute is this?

6-4 Skills Practice

Proportional and Nonproportional Relationships

Determine whether the set of numbers in each table is proportional. Explain.

1.
Number of Socks	1	2	3	4
Cost	$2	$4	$6	$6

2.
Number of Guests	2	4	6	8
Cookies	4	8	12	16

3.
Days	1	3	5	6
Pages Read	100	300	550	600

4.
Cups of Flour	2	4	8	10
Loaves of Bread	1	2	4	5

For Exercises 5 and 6, complete each table. Determine whether the pattern forms a proportion.

5. **BABY-SITTING** Aliya earns $7 per hour baby-sitting her neighbors.

Hours	1			
Earnings	$7			

6. **PIZZA** Antonio's Pizzaria charges $10 for a large pizza, plus $1.50 for each additional topping.

Number of Toppings	1				
Cost					

7. **TRAVELING** On a cross-country road trip, a family drives 240 miles each day. Write and solve an equation to determine how far the family has traveled after 4 days.

6-4 Practice

Proportional and Nonproportional Relationships

Determine whether the set of numbers in each table is proportional. Explain.

1.

Cups of Rice	1	2	2.5	3
Cups of Water	1.5	3	3.75	4.5

2.

Miles driven	1	2	6	9
Toll fare	$1.07	$1.14	$1.42	$1.63

For Exercises 3 and 4, write and solve an equation.

3. **JOBS** Sharif started a new job working 15 hours a week. After how many weeks will he have worked a total of 75 hours?

4. **GARDENING** During its first 50 days of growth, a sunflower grows about 4 cm per day. Using this rate, after how many days will a sunflower be 60 cm tall?

For Exercises 5–6, complete each table. Determine whether the pattern forms a proportion.

5. **TEXT MESSAGING** It costs Victoria $0.10 to send a text message.

Number of Messages	4				
Cost					

6. **WATER CONSUMPTION** Water flows out of a kitchen faucet at about 1.5 gallons per minute.

Minutes	0.5				
Gallons of Water					

7. **COOKING** The amount of time it takes to cook a turkey increases with the weight of the turkey. It is recommended that you cook a 10-lb turkey for 3 hours. An extra 12 minutes of cooking time is necessary for each additional pound of turkey. Is the cooking time proportional to the weight of the turkey? Explain your reasoning.

6-5 Skills Practice

Solving Proportions

Determine whether each pair of ratios forms a proportion.

1. $\dfrac{1}{5}, \dfrac{4}{20}$

2. $\dfrac{3}{8}, \dfrac{12}{32}$

3. $\dfrac{4}{5}, \dfrac{9}{10}$

4. $\dfrac{12}{20}, \dfrac{18}{30}$

5. $\dfrac{3}{4}, \dfrac{27}{36}$

6. $\dfrac{10}{18}, \dfrac{2}{9}$

7. $\dfrac{4}{9}, \dfrac{2}{3}$

8. $\dfrac{15}{18}, \dfrac{10}{12}$

9. $\dfrac{15}{24}, \dfrac{3}{8}$

10. $\dfrac{36}{72}, \dfrac{50}{100}$

11. $\dfrac{10}{8.4}, \dfrac{5}{4.2}$

12. $\dfrac{12}{4.8}, \dfrac{9}{3.2}$

ALGEBRA Solve each proportion.

13. $\dfrac{8}{4} = \dfrac{t}{8}$

14. $\dfrac{n}{9} = \dfrac{4}{18}$

15. $\dfrac{3}{v} = \dfrac{12}{32}$

16. $\dfrac{25}{60} = \dfrac{s}{12}$

17. $\dfrac{21}{28} = \dfrac{3}{w}$

18. $\dfrac{c}{12} = \dfrac{5}{6}$

19. $\dfrac{4}{r} = \dfrac{5}{20}$

20. $\dfrac{12}{18} = \dfrac{m}{81}$

21. $\dfrac{2}{9} = \dfrac{6}{k}$

22. $\dfrac{h}{35} = \dfrac{3}{7}$

23. $\dfrac{3}{16} = \dfrac{u}{40}$

24. $\dfrac{6}{a} = \dfrac{1}{3}$

25. $\dfrac{e}{9.5} = \dfrac{6.4}{7.6}$

26. $\dfrac{2.7}{3.0} = \dfrac{3.6}{x}$

27. $\dfrac{1.68}{w} = \dfrac{7}{12}$

6-5 Practice

Solving Proportions

Determine whether each pair of ratios forms a proportion.

1. $\frac{5}{8}, \frac{20}{32}$

2. $\frac{12}{28}, \frac{27}{63}$

3. $\frac{8}{50}, \frac{1}{43}$

4. $\frac{40}{48}, \frac{56}{42}$

5. $\frac{6.4}{16}, \frac{32}{80}$

6. $\frac{12}{18}, \frac{90}{135}$

7. $\frac{21}{24}, \frac{56}{64}$

8. $\frac{9}{16}, \frac{3}{4}$

9. $\frac{12}{32}, \frac{8}{3}$

10. $\frac{2.6}{4}, \frac{4.6}{8}$

11. $\frac{5.1}{1.7}, \frac{7.5}{2.5}$

12. $\frac{8.5}{25}, \frac{17}{50}$

ALGEBRA Solve each proportion.

13. $\frac{n}{12} = \frac{6}{18}$

14. $\frac{8}{v} = \frac{56}{105}$

15. $\frac{15}{35} = \frac{s}{7}$

16. $\frac{24}{30} = \frac{8}{w}$

17. $\frac{c}{28} = \frac{5}{7}$

18. $\frac{3}{r} = \frac{39}{65}$

19. $\frac{9}{15} = \frac{m}{25}$

20. $\frac{7.5}{6.0} = \frac{3.6}{x}$

21. $\frac{12}{25} = \frac{u}{40}$

22. $\frac{1}{a} = \frac{33}{132}$

23. $\frac{f}{5} = \frac{16}{40}$

24. $\frac{r}{6.5} = \frac{0.2}{1.3}$

25. $\frac{30}{14} = \frac{k}{1.54}$

26. $\frac{3.5}{7.2} = \frac{k}{57.6}$

27. $\frac{2.1}{42} = \frac{7}{t}$

28. **FOOD** Gayle is making fruit punch that consists of 2 quarts of juice and 1 quart of soda water. How much soda water does she need if she has 5 quarts of juice?

6-6 | Skills Practice

Scale Drawings and Models

On a set of architectural drawings for a new school building, the scale is $\frac{1}{4}$ inch = 2 feet. Find the missing lengths of the rooms.

	Room	Drawing Length	Actual Length
1.	Lobby		16 feet
2.	Principal's Office	1.25 inches	
3.	Library		20 feet
4.	School Room	3 inches	
5.	Science Lab	1.5 inches	
6.	Cafeteria		48 feet
7.	Music Room	4 inches	
8.	Gymnasium	13 inches	
9.	Auditorium		56 feet
10.	Teachers' Lounge	1.75 inches	

11. Refer to Exercises 1–10. What is the scale factor?

12. What is the scale factor if the scale is 10 inches = 1 foot?

13. **STRUCTURES** A barn is 40 feet wide by 100 feet long. Make a scale drawing of the barn that has a scale of $\frac{1}{2}$ inch = 10 feet.

14. **MAPS** On a map, the key indicates that 1 centimeter equals 3.5 meters. A road is shown on this map that runs for 30 centimeters. How long is this road?

6-6 Practice

Scale Drawings and Models

On a map, the scale is 5 centimeters = 2 kilometers. Find the missing distances.

	Location	Map Distance	Actual Distance
1.	Town A to Town B	10 cm	
2.	Town A to Town C		10 km
3.	Town A to Town D		5.6 km
4.	Town A to Town E	2 cm	
5.	Town A to Town F	0.5 cm	
6.	Town A to Town G		3.2 km
7.	Town A to Town H	0.25 cm	
8.	Town A to Town I		2.4 km
9.	Town A to Town J		0.04 km
10.	Town A to Town K	1 cm	
11.	Town A to Town L	2.5 cm	
12.	Town A to Town M		0.48 km

13. Refer to Exercises 1–12. What is the scale factor?

14. What is the scale factor if the scale is 15 inches = 1 yard?

15. STRUCTURES A barn is 50 feet wide by 80 feet long. Make a scale drawing of the barn that has a scale of $\frac{1}{2}$ inch = 10 feet.

16. PHOTOGRAPHY A man in a photograph is 1.5 inches in height. If the man is 6 feet tall, what is the scale?

6-7 Skills Practice

Similar Figures

In Exercises 1–10, the figures are similar. Find each missing measure.

1.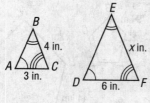

2.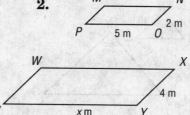

3.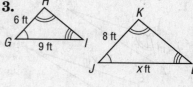

4.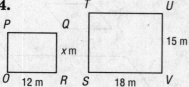

5.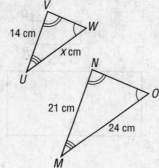

6.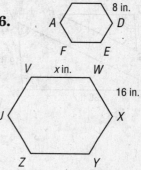

7.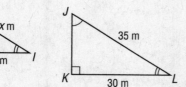

8.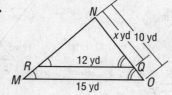

9. How far is the store from the bank?

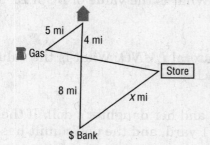

10. How far is the tree from the flagpole?

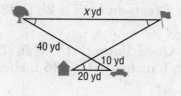

6-7 Practice

Similar Figures

In Exercises 1–8, the figures are similar. Find each missing measure.

1.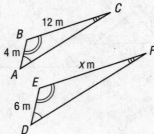

2.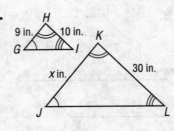

3.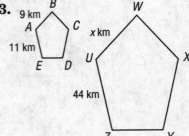

4.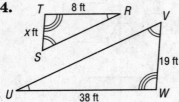

5.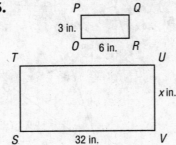

6.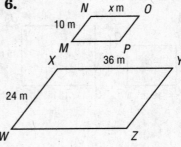

7.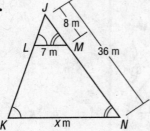

8.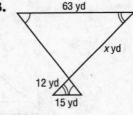

9. **GEOMETRY** Triangle ABC is similar to triangle DEF. What is the value of $\overline{BC}$ if $\overline{EF}$ is 36 feet, $\overline{AC}$ is 7 feet, and $\overline{DF}$ is 28 feet?

10. **GEOMETRY** Quadrilateral $RSTU$ is similar to quadrilateral $LMNO$. What is the value of $\overline{LO}$ if $\overline{RU}$ is 6 inches, $\overline{LM}$ is 45 inches, and $\overline{RS}$ is 9 inches?

11. **QUILTS** A woman sews similar quilts for her daughter and her daughter's doll. If the daughter's quilt has a length of 2 yards and a width of 1 yard, and the doll's quilt has a length of $\frac{1}{2}$ yard, what is the width of the doll's quilt?

6-8 Skills Practice

Dilations

Find the vertices of each figure after a dilation with the given scale factor k.
Then graph the image.

1. $k = 2$

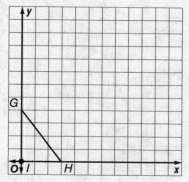

2. $k = 3$

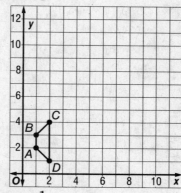

3. $k = \frac{1}{2}$

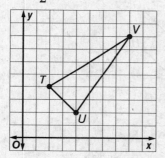

4. $k = \frac{1}{4}$

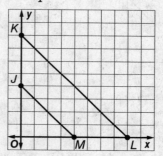

5. Find the vertices of figure $WXYZ$ after a dilation with a scale factor of $\frac{1}{3}$ if it
has vertices $W(-3, 6)$, $X(3, -3)$, $Y(-3, -6)$, and $Z(-6, -3)$. Then graph the image.

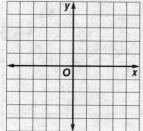

6. Find the scale factor for the dilation
shown at the right.

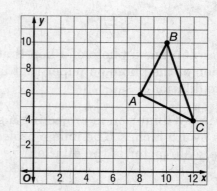

6-8 Practice

Dilations

Find the vertices of each figure after a dilation with the given scale factor k. Then graph the image.

1. $k = 3$

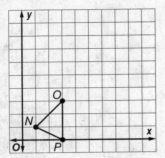

2. $k = 2.5$

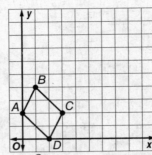

3. $k = \frac{1}{5}$

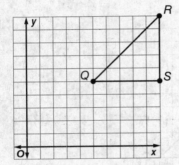

4. $k = \frac{2}{3}$

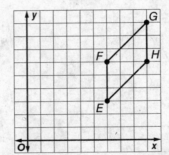

5. Find the vertices of figure $STUV$ after a dilation with a scale factor of 1.5 if it has vertices $S(-4, 1)$, $T(-4, 6)$, $U(-2, 8)$, and $V(-2, 3)$. Then graph the image.

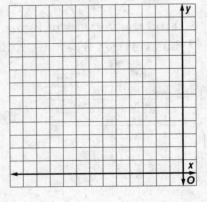

6. PHOTOS Jordan has a photo of a lion that is 4 inches by 6 inches. He wants to sketch a larger version of the photo on paper that is 14 inches by 21 inches. What is the scale factor of the dilation?

7. IMAGES Mrs. Williamson is projecting a slide on the wall. The image on the slide is 1.25 inches by 1.5 inches. The image projected on the wall is 20 inches by 24 inches. What is the scale factor of the dilation?

6-9 Skills Practice

Indirect Measurement

1. **ANIMALS** At the same time a 12-foot adult elephant casts a 4.8-foot shadow, a baby elephant casts a 2-foot shadow. How tall is the baby elephant?

2. **AIRPORTS** If a 12-meter-tall airplane hangar casts a 18-meter shadow at the same time a parked jet casts a 6-meter shadow, how tall is the jet?

3. **FERRIS WHEEL** Suppose a Ferris wheel is 160 feet high and casts a shadow that is 64 feet long. At the same time, a ticket booth next to the Ferris wheel casts a shadow that is 2.8 feet long. What is the height of the ticket booth?

4. **BUILDINGS** A building casts a shadow that is 72 feet long. A garage next to the building is 27 feet high and casts a shadow that is 4.5 feet long. What is the height of the building?

5. **FARMS** A silo casts a shadow that is 99 feet long. Next to the silo is an 18-foot-tall barn that casts a shadow that is 13.5 feet long. How tall is the silo?

6. **STATUES** In New Salem, North Dakota, there is a 38-foot-tall statue of a cow named Salem Sue. Suppose the statue's shadow was 57 feet long and a 3.5-foot child was standing next to the statue. How long would the child's shadow be?

7. **DISTANCES** The triangles below are similar. How far is the store from the bank?

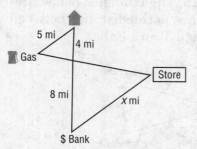

8. **GEOMETRY** The triangles below are similar. What is the value of x?

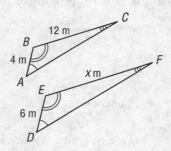

9. **MAPS** The triangles below are similar. How far is Clayton from Wiley's Junction?

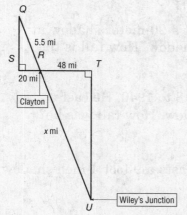

10. **DISTANCES** The triangles below are similar. What is the distance between the skate park and the movie theater?

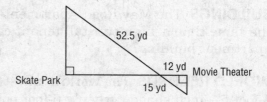

6-9 Practice

Indirect Measurement

1. GEOMETRY The triangles below are similar. What is the value of x?

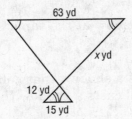

2. CANYONS In the figure, $\triangle CDE \sim \triangle GDF$. Find the distance across Rancher Canyon.

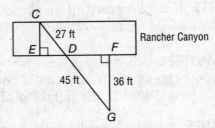

3. DISTANCES The triangles below are similar. How far is Dora's house from Micala's house?

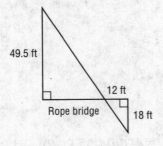

4. OUTDOOR SPORTS In the figure, $\triangle ABC \sim \triangle DBE$. How far is the archery range from the soccer field?

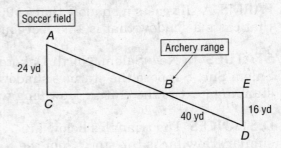

5. BRIDGES The triangles below are similar. How long is the rope bridge?

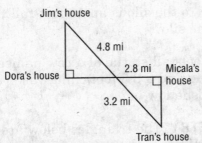

6. DISTANCES The triangles below are similar. What is the distance between Tarryhill and Tom's Falls?

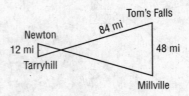

7. CHIMNEYS A 6-ft observer casts a 4-ft shadow at the same time a chimney casts a 238-foot shadow. How tall is the chimney?

8. BUILDINGS The May Road Apartments in Hong Kong cast a 90-meter shadow at the same time a 1.5-meter tall tenant casts a 0.75-meter shadow. How tall is the apartment building?

9. WORLD RECORDS The world's tallest man lived from 1918 to 1940. He cast a 4-foot $5\frac{1}{2}$-inch shadow when a 6-foot pole cast a 3-foot shadow. How tall was he?

10. SHADOWS A man casts a 14-foot shadow. A 4-foot child casts a 9-foot 4-inch shadow at the same time. How tall is the man?

7-1 Skills Practice

Fractions and Percents

Write each percent as a fraction or mixed number in simplest form.

1. 55%

2. 2%

3. $5\frac{1}{2}\%$

4. 30%

5. 300%

6. 12%

7. 50%

8. 90%

9. 85%

10. 28.2%

11. 0.25%

12. 0.2%

13. 7.5%

14. 6%

15. 10%

16. 275%

Write each fraction as a percent. Round to the nearest hundredth.

17. $\frac{3}{5}$

18. $\frac{11}{20}$

19. $\frac{1}{4}$

20. $\frac{5}{8}$

21. $\frac{23}{4}$

22. $\frac{4}{5}$

23. $\frac{3}{25}$

24. $\frac{7}{3}$

25. $2\frac{3}{10}$

26. $\frac{1}{6}$

27. $\frac{3}{4}$

28. $\frac{9}{10}$

29. $\frac{23}{50}$

30. $\frac{3}{8}$

31. $\frac{7}{5}$

32. $\frac{5}{4}$

7-1 Practice

Fractions and Percents

Write each percent as a fraction or mixed number in simplest form.

1. 35% 2. $8\frac{5}{6}\%$ 3. $10\frac{1}{2}\%$ 4. 8.4%

5. 500% 6. 32% 7. 80% 8. $\frac{1}{8}\%$

9. 65% 10. 48.5% 11. 0.15% 12. 0.9%

13. 2.5% 14. $25\frac{1}{3}\%$ 15. $\frac{1}{20}\%$ 16. 820%

Write each fraction as a percent. Round to the nearest hundredth.

17. $\frac{4}{15}$ 18. $\frac{3}{8}$ 19. $\frac{7}{9}$ 20. $\frac{5}{7}$

21. $4\frac{3}{4}$ 22. $\frac{300}{630}$ 23. $\frac{33}{40}$ 24. $\frac{9}{32}$

25. $\frac{11}{4}$ 26. $\frac{35}{8}$ 27. $\frac{1}{90}$ 28. $\frac{14}{25}$

29. $\frac{4}{11}$ 30. $\frac{5}{79}$ 31. $\frac{25}{8}$ 32. $2\frac{4}{13}$

33. **RIVERS** One of the longest rivers in the world is the Amazon river in South America. It is about 160% as long as the longest river in the United States, the Missouri River. Write 160% as a mixed number in simplest form.

34. **ZOOLOGY** A zoologist is tracking the number of baby animals born over the weekend at the zoo. Out of twenty new baby animals, 3 were antelopes. What percent of the baby animals born were antelopes?

7-2 Skills Practice

Fractions, Decimals, and Percents

Write each percent as a decimal.

1. 85%   2. 4%   3. 325%   4. 9.5%

5. 0.6%   6. 700%   7. $13\frac{1}{2}$%   8. 42.8%

9. 95%   10. 0.08%   11. 30%   12. 514%

13. 44.4%   14. 62%   15. 100%   16. 0.5%

Express each decimal or fraction as a percent. Round to the nearest tenth, if necessary.

17. 0.65   18. 0.772   19. 0.6   20. 3.45

21. 0.47   22. 0.01   23. 22.6   24. 0.79

25. 0.28   26. 0.355   27. 0.0015   28. 44

29. $\frac{11}{20}$ 30. $\frac{1}{4}$ 31. $\frac{5}{8}$ 32. $\frac{7}{5}$

33. $\frac{23}{4}$ 34. $\frac{4}{5}$ 35. $\frac{3}{25}$ 36. $\frac{7}{3}$

37. $2\frac{3}{10}$ 38. $\frac{1}{6}$ 39. $\frac{300}{630}$ 40. $\frac{9}{10}$

7-2 Practice

Fractions, Decimals, and Percents

Write each percent as decimal.

1. 17%

2. $6\frac{2}{3}\%$

3. $11\frac{3}{4}\%$

4. 9.2%

5. 800%

6. 43%

7. 20%

8. $\frac{3}{8}\%$

9. 75%

10. 26.3%

11. 0.12%

12. 0.01%

13. 5.5%

14. $68\frac{4}{5}\%$

15. $\frac{17}{20}\%$

16. 480%

Express each decimal or fraction as a percent. Round to the nearest tenth, if necessary.

17. 0.95

18. 0.255

19. 0.7

20. 8.75

21. 0.0048

22. 0.06

23. 19.8

24. 71

25. $\frac{33}{40}$

26. $\frac{9}{32}$

27. $\frac{3}{8}$

28. $\frac{11}{4}$

29. $\frac{35}{8}$

30. $\frac{1}{5}$

31. $\frac{14}{25}$

32. $\frac{4}{11}$

33. **SURVEYS** In a survey, 44% of the people said they voted for Mr. Johnson, while $\frac{2}{5}$ of the people said they voted for Ms. Smith. Which group is larger? Explain.

34. **LAUNDRY** Ben is sorting his clothes to be washed. Twenty-two percent of his dirty clothes are whites, $\frac{7}{20}$ are darks, and 0.45 are lights. Which group will have the least amount of laundry?

7-3 Skills Practice

Using the Percent Proportion

Use the percent proportion to solve each problem. Round to the nearest tenth, if necessary.

1. 64 is what percent of 200?

2. What percent of 12 is 9?

3. 2 is what percent of 80?

4. What percent of 42 is 32?

5. 10 is what percent of 60?

6. What percent of 30 is 6?

7. 15 is what percent of 24?

8. What percent of 36 is 9?

9. 28 is what percent of 42?

10. What percent of 72 is 21?

11. 8 is 40% of what number?

12. 16 is 5% of what number?

13. 25 is 80% of what number?

14. 0.84 is 28% of what number?

15. 71 is 10% of what number?

16. 52 is 97% of what number?

17. 39 is 17% of what number?

18. 12 is 4% of what number?

19. 48.5 is 7% of what number?

20. What is 10.6% of 11?

21. What is 15% of 98.4?

22. What is 0.5% of 75?

23. What is 4% of 512.5?

24. What is 50% of 1?

25. What is 25% of 12?

26. What is 12% of 25?

27. What is 90% of 50?

28. What is 50% of 90?

7-3 Practice

Using the Percent Proportion

Use the percent proportion to solve each problem. Round to the nearest tenth, if necessary.

1. 128 is what percent of 640?

2. What percent of 21 is 28?

3. 3.4 is what percent of 5?

4. What percent of 930 is 720?

5. 15 is what percent of 120?

6. What percent of 24 is 21?

7. 36 is what percent of 40?

8. What percent of 48 is 0.6?

9. 12 is 80% of what number?

10. 15 is 4% of what number?

11. 33 is 90% of what number?

12. 0.24 is 36% of what number?

13. 19 is 10% of what number?

14. 49 is 77% of what number?

15. 42 is 7.5% of what number?

16. 65 is 5% of what number?

17. 27.5 is 2% of what number?

18. What is 15.8% of 21?

19. What is 65% of 441.1?

20. What is 0.4% of 82?

21. What is 7% of 329.8?

22. What is 88% of 1?

23. SAVINGS Stacia has saved $36 toward the purchase of a new MP3 player. This is 28% of the total price. What is the price of the MP3 player?

24. PAINT About 42% of a paint mix is white. A painter orders 18 gallons of the paint mix. How much of it is white?

7-4 Skills Practice

Find Percent of a Number Mentally

Find the percent of each number mentally.

1. 10% of 582

2. 50% of 86

3. 40% of 1500

4. 20% of 75

5. 15% of 20

6. 80% of 45

7. 30% of 120

8. 75% of 44

9. 5% of 40

10. $33\frac{1}{3}$% of 99

11. 60% of 450

12. $37\frac{1}{2}$% of 56

13. 25% of 480

14. 300% of 5

15. 150% of 82

16. $66\frac{2}{3}$% of 210

17. 125% of 800

18. 175% of 400

Estimate.

19. 28% of 19

20. 55% of 32

21. 87% of 158

22. 35% of 544

23. 42% of 495

24. 19% of 319

25. 65% of 73

26. 8% of 224

27. 83% of 9

28. 17% of 331

29. 78% of 14

30. 12% of 879

31. $\frac{1}{3}$% of 941

32. $\frac{1}{2}$% of 376

33. $\frac{1}{5}$% of 2052

34. 164% of 318

35. 247% of 192

36. 508% of 1073

7-4 Practice

Find Percent of a Number Mentally

Find the percent of each number mentally.

1. 10% of 812

2. 50% of 1044

3. 40% of 25

4. 20% of 45

5. $62\frac{1}{2}$% of 80

6. 80% of 15

7. 30% of 400

8. 75% of 880

9. $16\frac{2}{3}$% of 72

10. $33\frac{1}{3}$% of 150

11. 60% of 2500

12. $37\frac{1}{2}$% of 48

13. 25% of 244

14. 900% of 3

15. 150% of 260

Estimate.

16. 31% of 62

17. 65% of 83

18. 87% of 850

19. 32% of 26

20. 47% of 213

21. 22% of 536

22. 68% of 12

23. 11% of 29

24. 78% of 4

25. $\frac{1}{2}$% of 381

26. $\frac{1}{6}$% of 567

27. $\frac{2}{3}$% of 856

28. 210% of 425

29. 153% of 801

30. 689% of 2981

31. MONEY Last week a waitress made $204 in tips. This week she made 135% of that. About how much did she make this week?

7-5 Skills Practice

Using Percent Equations

Solve each problem using a percent equation.

1. What is 5% of 80?

2. What is 10% of 100?

3. What is 58% of 35?

4. What is 32% of 150?

5. What is 91% of 3800?

6. Find 25% of 68.

7. Find 80% of 75.

8. Find 75% of 80.

9. Find 1.5% of 8400.

10. Find 33.5% of 22.

11. 23 is what percent of 115?

12. 27 is what percent of 75?

13. 80 is what percent of 160?

14. 85 is what percent of 500?

15. 48 is what percent of 30?

16. 321.3 is what percent of 918?

17. 0.6 is what percent of 2?

18. 126 is what percent of 140?

19. 21 is what percent of 1050?

20. 78 is what percent of 40?

21. 29 is 50% of what number?

22. 9 is 45% of what number?

23. 16 is 4% of what number?

24. 336 is 48% of what number?

25. 52 is 25% of what number?

26. 99 is 90% of what number?

27. 343 is 70% of what number?

28. 57 is 1% of what number?

29. 193.6 is 32% of what number?

30. 87.1 is 67% of what number?

7-5 Practice

Using Percent Equations

Solve each problem using a percent equation.

1. What is 5% of 224?

2. What is 18% of 65?

3. What is 63% of 300?

4. What is 40% of 980?

5. What is 18% of 650?

6. Find 2% of 820.

7. Find 75% of 312.

8. Find 312% of 75.

9. Find 5.6% of 1050.

10. Find 21.4% of 855.

11. 52.3 is what percent of 1046?

12. 48 is what percent of 75?

13. 100 is what percent of 250?

14. 96 is what percent of 400?

15. 10 is what percent of 625?

16. 49.8 is what percent of 415?

17. 0.4 is what percent of 5?

18. 157 is what percent of 2512?

19. 1206 is what percent of 8040?

20. 63 is what percent of 60?

21. 13 is 50% of what number?

22. 121 is 22% of what number?

23. 11 is 4% of what number?

24. 438 is 24% of what number?

25. 3570 is 42% of what number?

26. 8 is 1% of what number?

27. DINING Michael's bill at a restaurant was $46.32. He wants to leave a 17% tip. What will the new total be, including tip?

28. SHOPPING A jacket is on sale at 15% off the original price of $68.00. What is the sale price?

7-6 Skills Practice

Percent of Change

Find the percent of change. Round to the nearest tenth, if necessary. Then state whether the percent of change is an *increase* or *decrease*.

1. from 12 m to 18 m

2. from 27 days to 30 days

3. from $48.50 to $38.80

4. from 25 lb to 12 lb

5. from 10 mm to 3 mm

6. from $875 to $1000

7. from $18.10 to $22.50

8. from 32 people to 3040 people

9. from 28 stray cats to 5 stray cats

10. from 12 words to 90 words

11. from 47 mph to 35 mph

12. from 8 computers to 15 computers

13. from 34 workers to 28 workers

14. from 8056 snowflakes to 6381 snowflakes

Find the selling price for each item given the cost and the percent of markup.

15. necklace: $30; 25% markup

16. scooter: $15; 55% markup

17. tennis shoes: $8; 40% markup

18. computer: $200; 36% markup

19. watch: $22.50; 50% markup

20. video game: $40; 28% markup

21. jeans: $18; 33% markup

22. car: $16,000; 10% markup

7-6 Practice

Percent of Change

Find the percent of change. Round to the nearest tenth, if necessary. Then state whether the percent of change is an *increase* or *decrease*.

1. from 4 m to 5 m

2. from 75 minutes to 100 minutes

3. from $9.25 to $6.50

4. from 45 quarts to 8 quarts

5. from 21 mm to 13 mm

6. from $457 to $1000

7. from $39.50 to $40.00

8. from 9 students to 856 students

9. from 24 kittens to 7 kittens

10. from 15 songs to 105 songs

11. from 31 mph to 25 mph

12. from 4 paintings to 13 paintings

Find the selling price for each item given the cost and the percent of markup.

13. MP3 player: $150; 22% markup

14. sofa: $600; 31.5% markup

15. television: $669; 23% markup

16. dress: $16.25; 35% markup

17. speakers: $42; 27% markup

18. CD: $8.99; $34\frac{4}{5}$% markup

19. **VIDEOS** A video store is selling previously owned DVDs for 40% off the regular price of $15. What is the sale price of the DVDs?

20. **COOKIES** On Tuesday, a baker sold 132 cookies. On Wednesday, she sold 108 cookies. Find the percent of change to the nearest tenth of a percent.

7-7 Skills Practice

Simple and Compound Interest

Find the simple interest to the nearest cent.

1. $720 at 8% for 5 years

2. $385 at 6.2% for 3 years

3. $1200 at 4.25% for 18 months

4. $1950 at 7.5% for 6 months

5. $4250 at 9% for 10 years

6. $2008 at 6% for 3 months

7. $680 at 8% for 48 months

8. $1111 at 11% for 11 years

9. $1620 at 5.75% for 9 months

10. $800 at 12.5% for 2 years

11. $9500 at 3.3% for 30 months

12. $50 at 13.5% for 20 years

Find the total amount in each account to the nearest cent if the interest is compounded annually.

13. $2200 at 5% for 2 years

14. $3850 at 6.25% for 3 years

15. $4075 at 4.25% for 3 years

16. $325 at 7% for 6 years

17. $1000 at 12.25% for 4 years

18. $14,950 at 5.85% for 5 years

19. $750 at 12% for 2 years

20. $620 at 10.5% for 2 years

21. $4050 at 8.5% for 4 years

22. $1986 at 8.6% for 3 years

23. $11,300 at 9.1% for 3 years

24. $575 at 2.8% for 4 years

7-7 Practice

Simple and Compound Interest

Find the simple interest to the nearest cent.

1. $1300 at 6% for 7 years

2. $250 at 8% for 9 months

3. $725 at 3.25% for 6 months

4. $1900 at 5.5% for 36 months

5. $920 at 10.5% for 30 months

6. $1100 at 13% for 54 months

7. $550 at 5.75% for 4 years

8. $875 at 2.3% for 3 months

9. $22,800 at 9.3% for 33 months

10. $54,600 at 4.25% for 42 months

Find the total amount in each account to the nearest cent if the interest is compounded annually.

11. $450 at 5% for 3 years

12. $580 at 11.8% for 4 years

13. $6550 at 6.5% for 2 years

14. $2750 at 2.75% for 3 years

15. $1900 at 9% for 2 years

16. $13,900 at 12.5% for 5 years

17. $600 at 6% for 4 years

18. $2400 at 5.3% for 5 years

19. $64,000 at 3.25% for 3 years

20. $312,000 at 1.99% for 4 years

21. **INSTRUMENTS** Lane borrowed $1200 for a new drum set. She will be paying 6.5% in simple interest over the next 2 years. What is the total amount of interest she will be paying on the loan?

22. **SAVINGS** Luke puts $4800 in a savings account. He earns $16 each month for the next 60 months. Find the simple interest rate for his savings account.

23. **CARS** Toya has a car loan of $8500. Over the course of the loan, she paid a total of $5525 in interest at a rate of 13%. How many months was the car loan?

7-8 Skills Practice

Circle Graphs

Construct a circle graph for each set of data.

1.

School Night Bedtimes	
Time	Percent
8:00 P.M. – 8:59 P.M.	7
9:00 P.M. – 9:59 P.M.	50
10:00 P.M. – 10:59 P.M.	42
11:00 P.M. or later	1

2.

Number of Books Read over Summer Vacation	
Number	Percent
0	5
1	12
2	22
3	40
4	12
5 or more	9

3. **SAFETY** The circle graph at the right shows the results from a survey of 282 students ages 8–13 who were asked the question, "How often do you wear your bike helmet?" How many more students answered "Seldom or Never" than "Occasionally"?

How Often Do You Wear Your Bike Helmet?

4. **DOGS** The circle graph at the right shows the results from a survey of children of all ages about favorite breeds of dogs. Which two breeds were equally favored?

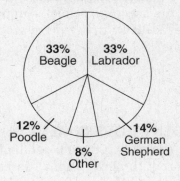

Favorite Dog Breeds

7-8 Practice

Circle Graphs

Construct a circle graph for each set of data.

1.

Allowance Paid to 12- to 17-year-olds	
Amount	**Percent**
None	29%
$1	2%
$2	2%
$3	2%
$4	0%
$5	15%
$6−10	23%
$11−20	19%
$21 and over	8%

Source: Nickelodeon

2.

How Frequently Do Parents Pay Allowance?	
Time Period	**Number**
Weekly	368
Biweekly	54
Monthly	49
Other (Usually when earned)	25

Source: Kids' Money

3. NEWS The circle graph at the right shows the results of a survey about how people get their news information. Suppose that 450 people were surveyed. How many more people watch the television news than read the paper as their primary source of news?

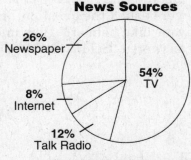

News Sources

26% Newspaper

54% TV

8% Internet

12% Talk Radio

4. TRANSPORTATION The circle graph at the right shows the results of a survey about how Middle School students travel to school. Suppose 300 students were surveyed. How many more students walk than take the bus?

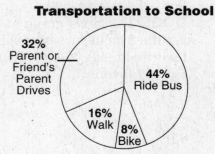

Transportation to School

32% Parent or Friend's Parent Drives

44% Ride Bus

16% Walk

8% Bike

5. STUDENT GOVERNMENT Jolene made a circle graph for her Government project to show the preference for student body president candidates. The central angle for the portion of the graph that represents Candidate A measures 32°. If 80 people prefer Candidate A, how many people did Jolene survey?

8-1 Skills Practice

Functions

Determine whether each relation is a function. Explain.

1. {(3,−8), (3, 2), (6, −1), (2, 2)}

2. {(0, 1), (−4, −3), (−3, 6), (3, 6)}

3. {(−6, 3), (2, −2), (0, 8), (1, 1)}

4. {(1, 8), (−6, 21), (−11, 21), (−3, 11), (0, 21)}

5.

x	1	−3	8	−8	20
y	2	6	6	5	11

6.

x	−1.2	1.1	1.7	−1.2	1.0
y	2.8	2.3	−2.4	2.3	2.6

7.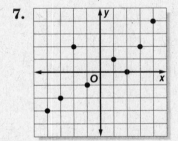

8.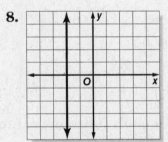

If $f(x) = 4x − 2$, find each function value.

9. $f(3)$ **10.** $f(9)$ **11.** $f(1)$ **12.** $f(4)$

13. $f(−2)$ **14.** $f(−10)$ **15.** $f(5)$ **16.** $f(−8)$

If $g(x) = 3x + 6$, find each function value.

17. $g(2)$ **18.** $g(7)$ **19.** $g(−4)$ **20.** $g(0)$

21. $g(−6)$ **22.** $g(−1)$ **23.** $g(9)$ **24.** $g(12)$

8-1 Practice

Functions

Determine whether each relation is a function. Explain.

1. {(4, −5), (0, −9), (1, 0), (7, 0)}

2. {(5, 2), (−2, 15), (−7, 15), (1, 5), (4, 15), (−7, 2)}

3.

x	−3.0	3.5	4.1	−3.0	3.4
y	4.2	3.7	−3.8	3.7	4.0

4.

x	7	14	11	−10	−1
y	−3	−9	−4	−3	15

5.

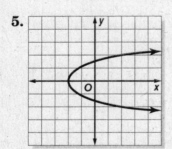

6.

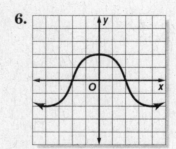

If $f(x) = \frac{1}{2}x + 5$, find each function value.

7. $f(24)$ **8.** $f(−30)$ **9.** $f(11)$ **10.** $f(−10)$

EMPLOYMENT For Exercises 11–14, use the table, which shows the percent of employed men and women in the U.S. labor force every five years from 1985 to 2005.

11. Is the relation (year, percent of men) a function? Explain.

12. Describe how the percent of employed men is related to the year.

Employed Members of Labor Force		
Year	Men (% of male population)	Women (% of female population)
1985	76.3	54.5
1990	76.4	57.5
1995	75.0	58.9
2000	78.9	67.3
2005	73.3	59.3

Source: U.S. Census Bureau

13. Is the relation (year, percent of women) a function? Explain.

14. Describe how the percent of employed women is related to the year.

8-2 Skills Practice

Sequences and Equations

Describe each sequence using words and symbols.

1. 7, 8, 9, 10, ...

2. 5, 6, 7, 8, ...

3. 7, 14, 21, 28, ...

4. 12, 24, 36, 48, ...

5. 3, 5, 7, 9, ...

6. 12, 21, 30, 39, ...

7. 55, 62, 69, 76, ...

8. 3, 21, 39, 57, ...

Write an equation that describes each sequence. Then find the indicated term.

9. 5, 8, 11, 14, ...; 9th term

10. 7, 16, 25, 34, ...; 15th term

11. 7, 9, 11, 13, ...; 18th term

12. 4, 10, 16, 22, ...; 10th term

13. 6, 17, 28, 39, ...; 8th term

14. 25, 44, 63, 82, ...; 12th term

15. 26, 29, 32, 35, ...; 14th term

16. 61, 83, 105, 127, ...; 20th term

8-2 Practice

Sequences and Equations

Describe each sequence using words and symbols.

1. 46, 52, 58, 64, …

2. 5, 13, 21, 29, …

3. 9, 14, 19, 24, …

4. 11, 14, 17, 20, …

5. 3, 5, 7, 9, …

6. 44, 60, 76, 92, …

Write an equation that describes each sequence. Then find the indicated term.

7. 20, 33, 46, 59, …; 17th term

8. 29, 38, 47, 56, …; 21st term

9. 101, 103, 105, 107, …; 30th term

10. 64, 67, 70, 73, …; 44th term

11. 26, 29, 32, 35, …; 57th term

12. 112, 140, 168, 196, …; 74th term

13. RUNNING Luisa ran 3 miles on the 3rd day of a month, and she repeated her run every 4 days for the rest of the month. What equation describes the sequence of days of that month that Luisa ran?

14. DEPRECIATION A new hybrid car costs $25,000. If it depreciates at $2000 of its value each year, find the value of the car over the next 5 years.

8-3 Skills Practice

Representing Linear Functions

Find four solutions of each equation. Write the solutions as ordered pairs.

1. $y = 8x - 4$

2. $y = -x + 12$

3. $4x - 4y = 24$

4. $x - y = -15$

5. $y = 7x - 6$

6. $y = -3x + 8$

7. $y = 12$

8. $4x - 2y = 0$

9. $4x - y = 4$

Graph each equation by plotting ordered pairs.

10. $y = 3x - 2$

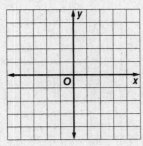

11. $y = -x + 3$

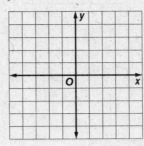

12. $y = -\dfrac{1}{2}x + \dfrac{3}{2}$

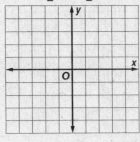

13. $y = -2x - 5$

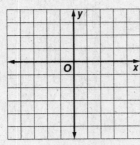

14. $y = 4x - 8$

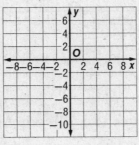

15. $y = \dfrac{2}{3}x - 2$

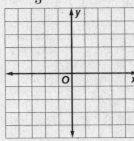

16. $y = -5x$

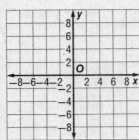

17. $y = -2x + 6$

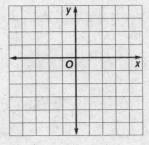

18. $y = 5x + 1$

8-3 Practice

Representing Linear Functions

Find four solutions of each equation. Write the solutions as ordered pairs.

1. $y = x - 5$

2. $y = -7$

3. $y = -3x + 1$

4. $x - y = 6$

5. $y = 2x + 4$

6. $7x - y = 14$

Graph each equation by plotting ordered pairs.

7. $y = 2x - 1$

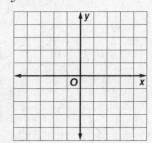

8. $y = -6x + 2$

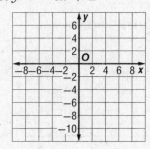

9. $y = x + 4$

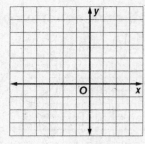

10. $y = 7$

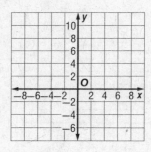

11. $y = 3x - 9$

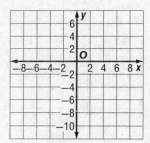

12. $y = \frac{1}{2}x - 6$

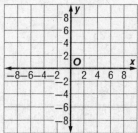

COOKING For Exercises 13–15, use the following information.

Kirsten is making gingerbread cookies using her grandmother's recipe and needs to convert grams to ounces. The equation $y = 0.04x$ describes the approximate number of ounces y in x grams.

13. Find three ordered pairs of values that satisfy this equation.

14. Draw the graph that contains these points.

15. Do negative values of x make sense in this case? Explain.

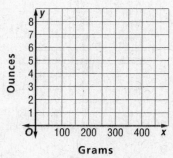

8-4 Skills Practice

Rate of Change

Find the rate of change for each linear function.

1.

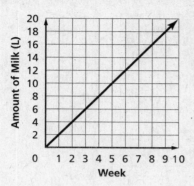

2.

Year	Salary ($)
x	y
1	21,000
2	23,500
3	26,000
4	28,500

3.

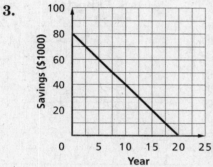

4.

Month	Number of Employees
x	y
0	0
2	22
4	44
6	66

5.

Time (min)	Temperature (°C)
x	y
0	9
1	23
2	37
3	51
4	65

6.

Price of Bulk Food

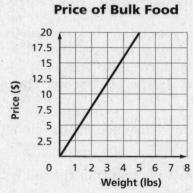

8-4 Practice

Rate of Change

Find the rate of change for each linear function.

1.
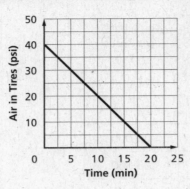

2.

Time (h)	Distance (km)
x	y
0	0
5	510
10	1020
15	1530

TRADE The graph shows the total U.S. exports from 1970 to 2005.

3. Find the approximate rate of change between 1970 and 1975.

4. Find the approximate rate of change between 2000 and 2005.

5. Between which two years was the rate of change the least?

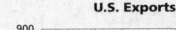

U.S. Exports

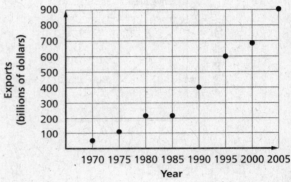

TRAFFIC MANAGEMENT For Exercises 6 and 7, use the following information.

San Diego reserves express lanes on the freeways for the use of carpoolers. In order to increase traffic flow during rush hours, other drivers may use the express lanes for a fee. The toll increases with the number of cars on the road. The table shows a sample of possible tolls.

6. Find the rate of change in the toll between 521 vehicles/h and 1122 vehicles/h.

7. Find the rate of change in the toll between 2204 vehicles/h and 1551 vehicles/h.

Toll ($)	Traffic Volume (vehicles/h)
1.00	521
2.00	1122
3.00	1551
4.00	2204

8-5 Skills Practice

Constant Rate of Change and Direct Variation

Find the constant rate of change for each linear function and interpret its meaning.

1.

Gallons	Quarts
x	y
1	4
2	8
3	12
4	16

2. **Allowance**

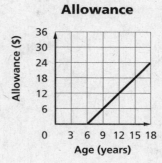

3. **Number of Teachers**

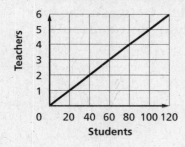

4.

Width (ft)	Height (in.)
x	y
2	10
4	14
6	18
8	22

Determine whether a proportional linear relationship exists between the two quantities shown in each of the functions indicated. Explain your reasoning.

5. Exercise 1

6. Exercise 2

7. Exercise 3

8. Exercise 4

8-5   Practice

Constant Rate of Change and Direct Variation

Find the constant rate of change for each linear function and interpret its meaning.

1.   **Fundraiser Profits**

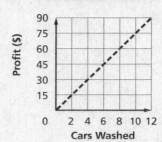

2.

Time (seconds)	Distance (yards)
x	y
1.2	6
2.4	8
3.6	10
4.8	12

Determine whether a proportional linear relationship exists between the two quantities shown in each of the functions indicated. Explain your reasoning.

3. Exercise 1

4. Exercise 2

PAPER COSTS The cost of paper varies directly with the number of reams bought. Suppose 2 reams cost $5.10.

5. Write an equation that could be used to find the cost of x reams of paper.

6. Find the cost of 15 reams of paper.

PHYSICAL SCIENCE Recall that the length a spring stretches varies directly with the amount of weight attached to it. A certain spring stretches 5 cm when a 10-gram weight is attached.

7. Write a direct variation equation relating the weight x and the amount of stretch y.

8. Estimate the stretch of the spring when it has a 42-gram weight attached.

8-6 Skills Practice

Slope

Find the slope of each line.

1.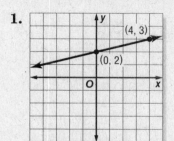

2.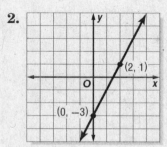

3.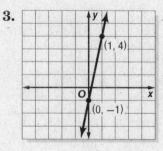

4.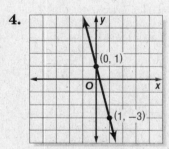

5.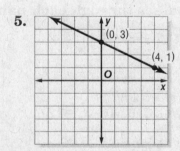

6.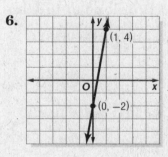

Find the slope of the line that passes through each pair of points.

7. $A(1, -5)$, $B(6, -7)$

8. $C(7, -3)$, $D(8, 1)$

9. $E(7, 2)$, $F(12, 6)$

10. $G(8, -3)$, $H(11, -2)$

11. $J(5, -9)$, $K(0, -12)$

12. $L(-4, 6)$, $M(5, 3)$

13. $P(2, -2)$, $Q(7, -1)$

14. $R(-5, -2)$, $S(-5, 3)$

15. $T(5, -6)$, $U(8, -12)$

16. $P(10, -2)$, $Q(3, -1)$

17. $R(6, -5)$, $S(7, 3)$

18. $T(1, 8)$, $U(7, 8)$

19. **CAMPING** A family camping in a national forest builds a temporary shelter with a tarp and a 4-foot pole. The bottom of the pole is even with the ground, and one corner is staked 5 feet from the bottom of the pole. What is the slope of the tarp from that corner to the top of the pole?

20. **ART** A rectangular painting on a gallery wall measures 7 meters high and 4 meters wide. What is the slope from the upper left corner to the lower right corner?

8-6 Practice

Slope

Find the slope of each line.

1.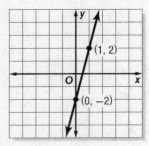

2.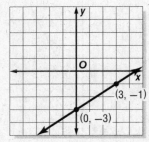

3.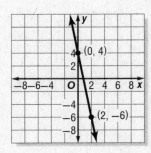

Find the slope of the line that passes through each pair of points.

4. $A(-10, 6)$, $B(-5, 8)$

5. $C(7, -3)$, $D(11, -4)$

6. $E(5, 2)$, $F(12, -3)$

7. $G(-15, 7)$, $H(-10, 6)$

8. $J(13, 0)$, $K(-3, -12)$

9. $L(-5, 3)$, $M(-4, 9)$

10. $P(12, 2)$, $Q(18, -2)$

11. $R(-2, -3)$, $S(-2, -5)$

12. $T(-13, 8)$, $U(21, 8)$

13. **CAKES** A wedding cake measures 2 feet high in the center and the diameter of the bottom tier is 12 inches. What is the slope of the cake?

14. **INSECTS** One particularly large ant hill found in 1997 measured 40 inches wide at the base and 18 inches high. What was the slope of the ant hill?

15. **ARCHAEOLOGY** Today, the Great Pyramid at Giza near Cairo, Egypt, stands 137 meters tall, coming to a point. Its base is a square with each side measuring 230 meters wide. What is the slope of the pyramid?

16. **BUSINESS** One warehouse uses 7-foot ramps to load its forklifts onto the flat beds of trucks for hauling. If the bed of a truck is 2 feet above the ground and the ramp is secured to the truck at its end, what is the slope of the ramp while in operation? Round to the nearest hundredth.

8-7 Skills Practice

Slope-Intercept Form

State the slope and the y-intercept for the graph of each equation.

1. $y = 12x - 4$

2. $y = \frac{1}{4}x + 3$

3. $3x - y = 6$

Given the slope and y-intercept, graph each line.

4. slope $= -2$,
   y-intercept $= 2$

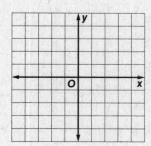

5. slope $= \frac{1}{2}$,
   y-intercept $= 4$

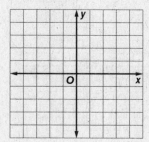

6. slope $= \frac{2}{3}$,
   y-intercept $= -3$

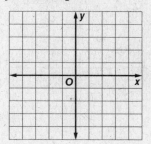

Graph each equation using the slope and y-intercept.

7. $y = 5x - 1$

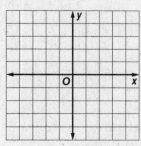

8. $y = \frac{1}{2}x + 2$

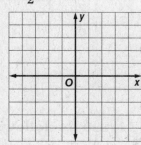

9. $y = -x + 2$

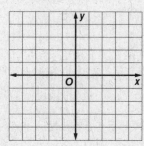

10. $y = 2x + 2$

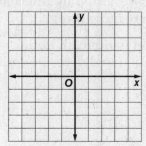

11. $y = -4x + 2$

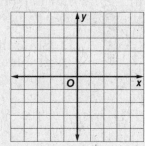

12. $y = x - 3$

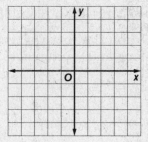

8-7 Practice

Slope-Intercept Form

State the slope and the *y*-intercept of the graph of each line.

1. $4x - y = 6$ **2.** $3x + 2y = 8$ **3.** $y - \frac{1}{2}x = \frac{3}{4}$

Graph each equation using the slope and *y*-intercept.

4. slope $= \frac{3}{4}$,
    y-intercept $= -3$

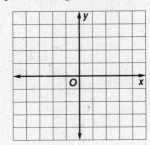

5. slope $= \frac{5}{6}$,
    y-intercept $= 1$

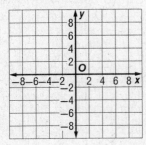

6. slope $= 1$,
    y-intercept $= 5$

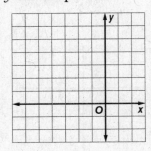

7. $y = -\frac{1}{2}x - 4$

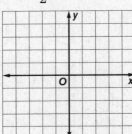

8. $y = x - 4$

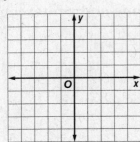

9. $y = -6x + 3$

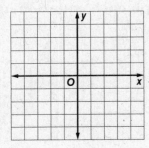

EXERCISE   **For Exercises 10 and 11, use the following information.**

A person weighing 150 pounds burns about 320 Calories per hour walking at a moderate pace. Suppose that the same person burns an average of 1500 Calories per day through basic activities. The total Calories *y* burned by that person can be represented by the equation $y = 320x + 1500$, where *x* represents the number of hours spent walking.

10. Graph the equation using the slope and *y*-intercept.

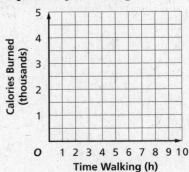

11. State the slope and *y*-intercept of the graph of the equation and describe what they represent.

8-8 Skills Practice

Writing Linear Equations

Write an equation for each line in slope-intercept form.

1. slope = 7,
y-intercept = 2

2. slope = −5,
y-intercept = −3

3. slope = $\frac{3}{5}$,
y-intercept = 6

4. slope = −6,
y-intercept = 7

5. slope = $\frac{2}{7}$,
y-intercept = 1

6. slope = $\frac{4}{3}$,
y-intercept = −4

7.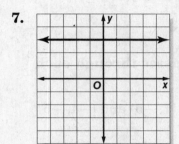

8.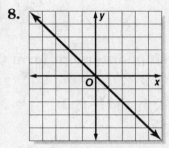

9.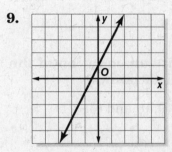

10.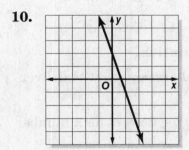

11.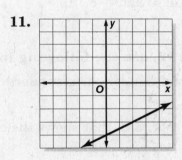

12.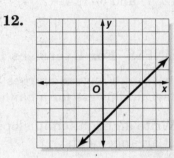

Write an equation of the line in slope-intercept that passes through each pair of points.

13. (9, −1) and (6, −2)

14. (12, 5) and (−4, 1)

15. (10, −6) and (−2, −6)

16. (4, 6) and (1, 3)

17. (6, 3) and (−6, 9)

18. (8, −4) and (−4, −1)

19. (5, 0) and (2, −3)

20. (12, −2) and (6, 2)

21. (−5, 10) and (3, −6)

8-8 Practice

Writing Linear Equations

Write an equation for each line in slope-intercept form.

1. slope = 3,
y-intercept = −2

2. slope = 0,
y-intercept = 7

3.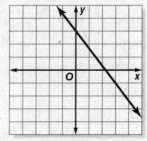

4.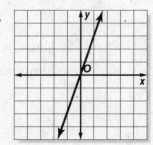

Write an equation of the line in slope-intercept form that passes through each pair of points.

5. (9, 0) and (6, −1)

6. (8, 6) and (−8, 2)

7. (7, −5) and (−4, −5)

8. (2, 7) and (−1, 4)

9. (4, 4) and (−8, 10)

10. (0, 2) and (−3, 14)

BUSINESS For Exercises 11 and 12, use the following information.

Flourishing Flowers charges $125 plus $60 for each standard floral arrangement to deliver and set up flowers for a banquet.

11. Write an equation in slope-intercept form that shows the cost y for flowers for x number of arrangements.

12. Find the cost of providing 20 floral arrangements.

INSULATION For Exercises 13 and 14, use the following information.

Renata González wants to increase the energy efficiency of her house by adding to the insulation previously installed. The better a material protects against heat loss, the higher its R-value, or resistance to heat flow. The table shows the R-value of fiberglass blanket insulation per inch of thickness. The existing insulation in Renata's attic has an R-value of 10.

R-value	Thickness (in.)
0.0	0
3.2	1
6.4	2
9.6	3

Source: Oak Ridge National Laboratory

13. Write an equation in slope-intercept form that shows the total R-value y in the attic if she adds x number of inches of additional insulation.

14. Estimate the total R-value in the attic if she adds 6 inches of insulation.

8-9 Skills Practice

Prediction Equations

CONSTRUCTION For Exercises 1 and 2, use the table that shows the average hourly wage of U.S. construction workers from 1980 to 2005.

Year	Average Hourly Earnings ($)
1980	9.94
1985	12.32
1990	13.77
1995	15.09
2000	17.48
2005	19.46

Source: *The New York Times Almanac*

1. Make a scatter plot and draw a line of fit for the data.

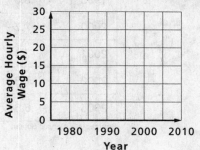

2. Use the line of fit to predict the average hourly wage of construction workers in 2010.

MINING For Exercises 3 and 4, use the table that shows the number of persons employed in mining from 1980 to 2005.

Year	Employees (thousands)
1980	1027
1985	927
1990	709
1995	581
2000	475
2005	318

Source: U.S. Census Bureau

3. Make a scatter plot and draw a line of fit for the data.

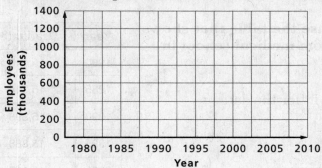

4. Write an equation for the line of fit and use it to predict the number of persons employed in mining in 2010.

8-9 Practice

Prediction Equations

BEVERAGES For Exercises 1 and 2, use the table that shows the amount of whole milk consumed per person in the United States.

Year	Gallons per Person
1990	10.2
1995	8.3
2000	7.7
2001	7.4
2002	7.3
2005	6.6

Source: U.S. Census Bureau

1. Make a scatter plot and draw a line of fit for the data.

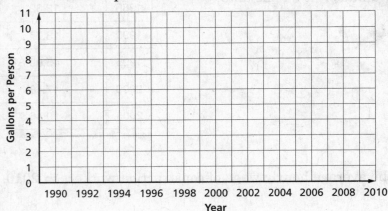

2. Use the line of fit to predict the amount of whole milk consumed per person in 2010.

EDUCATION For Exercises 3 and 4, use the table that shows the number of students graduating from medical school in the United States from 1980 to 2005.

Year	Graduates
1980	15,113
1985	16,318
1990	15,398
1995	15,888
2000	16,112
2005	16,110

Source: U.S. Census Bureau

3. Make a scatter plot and draw a line of fit for the data.

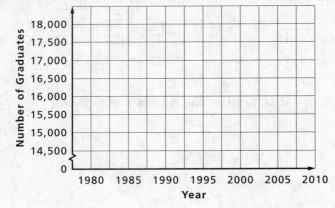

4. Write an equation for the line of fit and use it to predict the number of medical school graduates in 2010.

8-10 Skills Practice

Systems of Equations

Solve each system of equations by graphing.

1. $y = x - 9$

$y = 2x + 4$

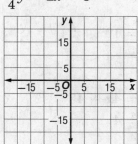

2. $y = -2x$

$y = x + 3$

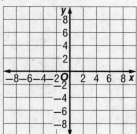

3. $\frac{1}{2}y = 4x - 6$

$\frac{1}{4}y = 2x - 3$

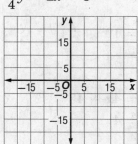

4. $y = -x$

$y = x + 6$

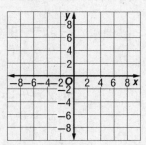

Solve each system of equations by substitution.

5. $y = x - 8$

$y = 1$

6. $y = x + 4$

$y = 0$

7. $y = x + 9$

$y = -4$

8. $y = 11 - x$

$y = -2$

9. $y = 3x + 10$

$x = 5$

10. $y = 2x$

$x = -4$

11. $y = -2x + 1$

$x = -3$

12. $y = 5 + 3x$

$y = -4$

13. $16 = 4x - y$

$y = 2x$

14. $26 = y + x$

$y = x$

8-10 Practice

Systems of Equations

Solve each system of equations by graphing.

1. $y = x + 3$

$y = 4x$

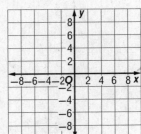

2. $y = x - 3$

$y = x + 3$

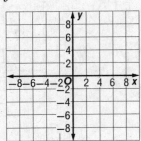

3. $4x + y = 18$

$y = -x$

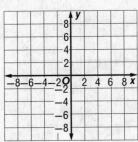

Solve each system of equations by substitution.

4. $y = x - 2$

$y = 4$

5. $y = 13 - x$

$y = -5$

6. $y = 10x + 24$

$y = -6$

7. $y = 5x + 12$

$y = -x$

8. $y = -2x$

$x = 0$

9. $y = 4x + 45$

$x = 4y$

10. CHOIR There are twice as many girls as boys in the school chorus. There are 8 fewer boys than girls in the chorus. Write a system of equations to represent this situation. Then solve the system by graphing. Explain what the solution means.

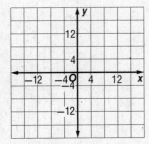

11. FOOD The cost of 8 muffins and 2 quarts of milk is $18. The cost of 3 muffins and 1 quart of milk is $7.50. Write a system of equations to represent this situation. Solve the system of equations by substitution. Explain what the solution means.

NAME _____ DATE _____ PERIOD _____

9-1 Skills Practice

Powers and Exponents

Write each expression using exponents.

1. $7 \cdot 7$

2. $(-3)(-3)(-3)(-3)(-3)$

3. 4

4. $(k \cdot k)(k \cdot k)(k \cdot k)$

5. $p \cdot p \cdot p \cdot p \cdot p$

6. $3 \cdot 3$

7. $(-a)(-a)(-a)(-a)$

8. $6 \cdot 6 \cdot 6 \cdot 6$

9. $9 \cdot 9 \cdot 9$

10. $4 \cdot y \cdot z \cdot z \cdot z$

11. $r \cdot r \cdot r \cdot r \cdot t \cdot u \cdot u$

12. $5 \cdot 5 \cdot 5 \cdot q \cdot q$

13. $8 \cdot 8 \cdot c \cdot c \cdot c \cdot c \cdot d \cdot d \cdot d$

14. $(-w)(-w)(v)(v)(v)(v)(v)$

15. $b \cdot b \cdot b \cdot b \cdot b \cdot b \cdot b \cdot b \cdot b$

16. $10 \cdot 10 \cdot 10 \cdot (-2) \cdot (-2) \cdot (-2) \cdot m \cdot m \cdot m$

Evaluate each expression if $a = -3$, $b = 8$, and $c = 2$.

17. 4^c

18. c^0

19. b^3

20. $c^3 \cdot 3^c$

21. 3^c

22. c^4

23. $c^2 + a$

24. $2b^2$

25. $b^2 + c^3$

26. a^2

27. a^3

28. $b^2 + a^3$

29. $b^2 a$

30. $(b - c)^2$

Copyright © Glencoe/McGraw-Hill, a division of The McGraw-Hill Companies, Inc.

9-1 Practice

Powers and Exponents

Write each expression using exponents.

1. $11 \cdot 11 \cdot 11$

2. $2 \cdot 2 \cdot 2 \cdot 2 \cdot 2 \cdot 2 \cdot 2 \cdot 2$

3. 5

4. $(-4)(-4)$

5. $a \cdot a \cdot a \cdot a$

6. $n \cdot n \cdot n \cdot n \cdot n$

7. $4 \cdot 4 \cdot 4$

8. $(b \cdot b)(b \cdot b)(b \cdot b)$

9. $(-v)(-v)(-v)(-v)$

10. $x \cdot x \cdot z \cdot z \cdot z$

11. $2 \cdot 2 \cdot 2 \cdot 2 \cdot 2 \cdot t \cdot t$

12. $m \cdot m \cdot m \cdot n \cdot p \cdot p$

13. $(-6)(-6)(-6)(-d)(-d)(-d)(-d)$

14. $3 \cdot 3 \cdot 3 \cdot 3 \cdot p \cdot q \cdot q \cdot q$

Evaluate each expression if $x = 3$, $y = -2$, and $z = 4$.

15. y^z

16. x^z

17. y^x

18. 51^0

19. z^2

20. x^2

21. 9^x

22. $z^2 \cdot 2^2$

23. y^5

24. $z^2 - y^4$

25. $x^2 + y^2 + z^2$

26. $z^2 - x^2$

FAMILY TREE **For Exercises 27 and 28, refer to the following information.**

When examining a family tree, the branches are many. You are generation "now." One generation ago, your 2 parents were born. Two generations ago, your 4 grandparents were born.

27. How many great-grandparents were born three generations ago?

28. How many "great" grandparents were born ten generations ago?

9-2 | Skills Practice

Prime Factorization

Determine whether each number is *prime* or *composite*.

1. 41

2. 29

3. 87

4. 36

5. 57

6. 61

7. 71

8. 103

9. 39

10. 91

11. 47

12. 67

Write the prime factorization of each number. Use exponents for repeated factors.

13. 20

14. 40

15. 32

16. 44

17. 90

18. 121

19. 46

20. 30

21. 65

22. 80

Factor each monomial.

23. $15t$

24. $16r^2$

25. $-11m^2$

26. $-49y^3$

27. $21ab$

28. $-42xyz$

29. $45j^2k$

30. $17u^2v^2$

31. $27d^4$

32. $-16cd^2$

9-2 Practice

Prime Factorization

Determine whether each number is *prime* or *composite*.

1. 11   **2.** 63

3. 73   **4.** 75

5. 49   **6.** 69

7. 53   **8.** 83

Write the prime factorization of each number. Use exponents for repeated factors.

9. 33   **10.** 24

11. 72   **12.** 276

13. 85   **14.** 1024

15. 95   **16.** 200

17. 243   **18.** 735

Factor each monomial.

19. $35v$ **20.** $49c^2$

21. $-14b^3$ **22.** $-81h^2$

23. $33wz$ **24.** $-56ghj$

25. NUMBER THEORY *Twin primes* are a pair of consecutive odd primes, which differ by 2. For example, 3 and 5 are twin primes. Find the twin primes less than 100. (*Hint*: There are 8 pairs of twins less than 100.)

9-3 Skills Practice

Multiplying and Dividing Monomials

Find each product or quotient. Express using exponents.

1. $2^3 \cdot 2^5$

2. $10^2 \cdot 10^7$

3. $1^4 \cdot 1$

4. $6^3 \cdot 6^3$

5. $(-3)^2(-3)^3$

6. $(-9)^2(-9)^2$

7. $a^2 \cdot a^3$

8. $n^8 \cdot n^3$

9. $(p^4)(p^4)$

10. $(z^6)(z^7)$

11. $(6b^3)(3b^4)$

12. $(-v)^3(-v)^7$

13. $11a^2 \cdot 3a^6$

14. $10t^2 \cdot 4t^{10}$

15. $(8c^2)(9c)$

16. $(4f^8)(5f^6)$

17. $\dfrac{5^{10}}{5^2}$

18. $\dfrac{10^6}{10^2}$

19. $\dfrac{7^9}{7^6}$

20. $\dfrac{12^8}{12^3}$

21. $\dfrac{100^9}{100^8}$

22. $\dfrac{(-2)^3}{-2}$

23. $\dfrac{r^8}{r^7}$

24. $\dfrac{z^{10}}{z^8}$

25. $\dfrac{q^8}{q^4}$

26. $\dfrac{g^{12}}{g^8}$

27. $\dfrac{(-y)^7}{(-y)^2}$

28. $\dfrac{(-z)^{12}}{(-z)^5}$

29. the product of two squared and two to the sixth power

30. the quotient of ten to the seventh power and ten cubed

31. the product of y squared and y cubed

32. the quotient of a to the twentieth power and a to the tenth power

9-3 Practice

Multiplying and Dividing Monomials

Find each product or quotient. Express using exponents.

1. $4^2 \cdot 4^3$

2. $9^8 \cdot 9^6$

3. $7^4 \cdot 7^2$

4. $13^2 \cdot 13^4$

5. $(-8)^5(-8)^3$

6. $(-21)^9(-21)^5$

7. $t^9 \cdot t^3$

8. $h^4 \cdot h^{13}$

9. $(m^6)(m^6)$

10. $(u^{11})(u^{10})$

11. $(-r)^7(-r)^{20}$

12. $(-w)(-w)^9$

13. $4d^5 \cdot 8d^6$

14. $7j^{50} \cdot 6j^{50}$

15. $-5b^9 \cdot 6b^2$

16. $12^1 \cdot 12^2$

17. $\dfrac{6^{11}}{6^3}$

18. $\dfrac{15^3}{15^2}$

19. $\dfrac{9^9}{9^7}$

20. $\dfrac{18^4}{18^4}$

21. $\dfrac{(-7)^6}{(-7)^5}$

22. $\dfrac{95^{21}}{95^{18}}$

23. $\dfrac{v^{30}}{v^{20}}$

24. $\dfrac{n^{19}}{n^{11}}$

25. the product of five cubed and five to the fourth power

26. the quotient of eighteen to the ninth power and eighteen squared

27. the product of z cubed and z cubed

28. the quotient of x to the fifth power and x cubed

29. **SOUND** Decibels are units used to measure sound. The softest sound that can be heard is rated as 0 decibels (or a relative loudness of 1). Ordinary conversation is rated at about 60 decibels (or a relative loudness of 10^6). A rock concert is rated at about 120 decibels (or a relative loudness of 10^{12}). How many times greater is the relative loudness of a rock concert than the relative loudness of ordinary conversation?

9-4 Skills Practice

Negative Exponents

Write each expression using a positive exponent.

1. 3^{-4}

2. 8^{-7}

3. 10^{-4}

4. $(-2)^{-6}$

5. $(-40)^{-3}$

6. $(-17)^{-12}$

7. n^{-10}

8. b^{-8}

9. q^{-5}

10. m^{-4}

11. v^{-11}

12. p^{-2}

Write each fraction as an expression using a negative exponent other than −1.

13. $\dfrac{1}{8^2}$

14. $\dfrac{1}{10^5}$

15. $\dfrac{1}{2^3}$

16. $\dfrac{1}{6^7}$

17. $\dfrac{1}{17^4}$

18. $\dfrac{1}{21^2}$

19. $\dfrac{1}{3^7}$

20. $\dfrac{1}{9^2}$

21. $\dfrac{1}{3^2}$

22. $\dfrac{1}{121}$

23. $\dfrac{1}{25}$

24. $\dfrac{1}{36}$

Evaluate each expression if $x = 1$, $y = 2$, and $z = -3$.

25. y^{-z}

26. z^{-2}

27. x^{-8}

28. y^{-5}

29. z^{-3}

30. y^{-1}

31. z^{-4}

32. 5^z

33. x^{-99}

34. 1^z

35. 4^z

36. y^z

9-4 Practice

Negative Exponents

Write each expression using a positive exponent.

1. 7^{-8}

2. 10^{-6}

3. 23^{-1}

4. $(-5)^{-2}$

5. $(-18)^{-10}$

6. m^{-99}

7. $(-1)^{-12}$

8. c^{-6}

9. p^{-5}

10. g^{-17}

11. $5z^{-4}$

12. $3t^{-1}$

Write each fraction as an expression using a negative exponent.

13. $\dfrac{1}{2^{10}}$

14. $\dfrac{1}{29^3}$

15. $\dfrac{1}{4^4}$

16. $\dfrac{1}{39}$

17. $\dfrac{1}{81^7}$

18. $\dfrac{1}{m^4}$

19. $\dfrac{1}{x^3}$

20. $\dfrac{1}{a^2}$

21. $\dfrac{1}{49}$

22. $\dfrac{1}{8}$

23. $\dfrac{1}{144}$

24. $\dfrac{1}{169}$

Evaluate each expression if $x = 3$, $y = -2$, and $z = 4$.

25. x^{-4}

26. y^{-2}

27. y^{-5}

28. z^{-4}

29. 5^y

30. 10^y

31. $3z^{-1}$

32. z^y

33. $(xz)^{-2}$

34. **HAIR** Hair grows at a rate of $\dfrac{1}{64}$ inch per day. Write this number using negative exponents.

9-5 Skills Practice

Scientific Notation

Express each number in standard form.

1. 1.5×10^3

2. 4.01×10^4

3. 6.78×10^2

4. 5.925×10^6

5. 7.0×10^8

6. 9.99×10^7

7. 3.0005×10^5

8. 2.54×10^5

9. 1.75×10^4

10. 1.2×10^{-6}

11. 7.0×10^{-1}

12. 6.3×10^{-3}

13. 5.83×10^{-2}

14. 8.075×10^{-4}

15. 1.1×10^{-5}

16. 7.3458×10^7

Express each number in scientific notation.

17. 1,000,000

18. 17,400

19. 500

20. 803,000

21. 0.00027

22. 5300

23. 18

24. 0.125

25. 17,000,000,000

26. 0.01

27. 21,800

28. 2,450,000

29. 0.0054

30. 0.000099

31. 8,888,800

32. 0.00912

Choose the greater number in each pair.

33. 8.8×10^3, 9.1×10^{-4}

34. 5.01×10^2, 5.02×10^{-1}

35. 6.4×10^3, 900

36. 1.9×10^{-2}, 0.02

37. 2.2×10^{-3}, 2.1×10^2

38. 8.4×10^2, 839

Order each set of numbers from least to greatest.

39. 3.6×10^4; 5.8×10^{-3}; 2.1×10^6; 3.5×10^5

40. 64,000,000; 6.2×10^8; 6,400,000; 6.4×10^5

9-5 Practice

Scientific Notation

Express each number in standard form.

1. 2.4×10^4

2. 9.0×10^3

3. 4.385×10^7

4. 1.03×10^8

5. 3.05×10^2

6. 5.11×10^{10}

7. 6.000032×10^6

8. 1.0×10^1

9. 8.75×10^5

10. 8.49×10^{-2}

11. 7.1×10^{-6}

12. 1.0×10^{-3}

13. 4.39×10^{-7}

14. 1.25×10^{-4}

Express each number in scientific notation.

15. 40,000

16. 16

17. 876,000,000

18. 4500

19. 151

20. 0.00037

21. 83,000,000

22. 919,100

23. 5,000,000,000,000

24. 0.13

25. 0.0000007

26. 0.0067

Order each set of numbers from least to greatest.

27. 7.35×10^4, 1.7×10^{-6}, 8.26×10^3, 9.3×10^{-2}

28. 0.00048, 4.37×10^{-4}, 4.02×10^{-3}, 0.04

NIAGARA FALLS **For Exercises 29 and 30, use the following information.**

Every minute, 840,000,000,000 drops of water flow over Niagara Falls.

29. Write this number in scientific notation.

30. How many drops flow over the falls in a day?

9-6 Skills Practice

Powers of Monomials

Simplify.

1. $(7^2)^9$

2. $(12^8)^5$

3. $(-15^6)^4$

4. $(h^3)^{-7}$

5. $(6^5)^{10}$

6. $(-f^7)^2$

7. $(k^6)^{-6}$

8. $(a^9)^{-3}$

9. $(16^2)^8$

10. $(42^5)^4$

11. $(m^3)^{-8}$

12. $(p^{-8})^7$

13. $(90^5)^9$

14. $(-3^6)^5$

15. $(-b^3)^4$

16. $(6y^6)^4$

17. $(12k^5)^2$

18. $(-7c^{-6})^3$

19. $(11n^5)^3$

20. $(3p^{10})^5$

21. $(-5g^{-8})^4$

22. $(4a^2b^8)^5$

23. $(-10h^5j^9)^6$

24. $(9r^2t^{-3})^2$

25. $(-8x^{-2}y^{-6})^3$

26. $(2g^2h^3)^8$

27. $(-3a^4b^3)^4$

28. $(20t^7u^{12})^3$

29. $(40x^5z^{-2})^2$

30. $(13f^9g^2)^2$

9-6 Practice

Powers of Monomials

Simplify.

1. $(19^3)^6$

2. $(-8^4)^9$

3. $(28^2)^{-5}$

4. $(q^8)^{-2}$

5. $(w^3)^4$

6. $(-46^{10})^7$

7. $(b^9)^{-3}$

8. $(m^5)^{-2}$

9. $(-103^4)^{12}$

10. $(88^3)^7$

11. $(x^{-2})^4$

12. $(v^{-4})^4$

13. $(7l^8)^2$

14. $(-4x^3)^4$

15. $(-9f^7)^3$

16. $(12r^{-5})^2$

17. $(3s^8)^4$

18. $(-5y^7)^4$

19. $(10u^5v^{-3})^5$

20. $(-2h^5i^3)^7$

21. $(4c^{-8}d^6)^2$

22. $(9f^5g^{-5})^3$

23. $(11j^{-4}k^{-2})^2$

24. $(-3b^5c^{-7})^5$

25. $(-5x^{-3}y^{-5}z^2)^2$

26. $(-7a^5b^6c^7)^2$

27. $(4g^6h^{-2}i^{-8})^6$

28. $[(4^2)^3]^2$

29. $(0.6c^{-5})^2$

30. $(\frac{1}{4}r^4s^6)^3$

31. GEOMETRY Find the area of a square with sides of length $6x^2y^5$ units.

32. GEOMETRY Find the volume of a cube with sides of length $4a^4b^6$ units.

33. ASTRONOMY The diameter of the Sun is 8.65×10^5 miles. Use the formula $A = 3.14 \cdot r^2$ to find the area of the cross section of the Sun at the equator.

9-7 Skills Practice

Linear and Nonlinear Functions

Determine whether each graph, equation, or table represents a *linear* or *nonlinear* function. Explain.

1.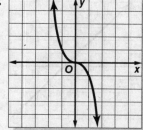

2.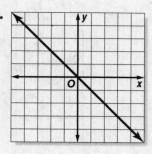

3.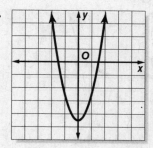

4. $y = \frac{x}{2} + 1$

5. $y = \frac{2}{x} + 10$

6. $y = 8x$

7. $y = 6$

8. $2x - y = 5$

9. $y = x^2 + 4$

10. $y + 4x^2 - 1 = 0$

11. $2y - 8x + 11 = 0$

12. $y = \sqrt{3x} - 2$

13.

x	y
1	8
2	5
3	2
4	−1

14.

x	y
6	1
12	3
18	6
24	10

15.

x	y
20	−4
15	−2
10	0
5	2

9-7 Practice

Linear and Nonlinear Functions

Determine whether each graph, equation, or table represents a *linear* or *nonlinear* function. Explain.

1.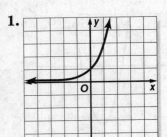

2.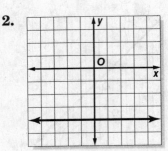

3.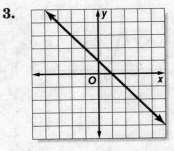

4. $5x - y = 15$

5. $3y + 12x^2 = 0$

6. $5y - x + 3 = 0$

7. $y = 6\sqrt{x} + 10$

8. $y = \dfrac{8}{x}$

9. $y = -x^2 + 2$

10.

x	y
1	1.0
2	0.8
3	0.6
4	0.4

11.

x	y
44	0
48	2.5
52	5.0
56	7.5

12.

x	y
3	1
6	−2
9	− 5
12	− 14

13. **GEOMETRY**   The graph shows how the area of a square increases as the perimeter increases. Is this relationship linear or nonlinear? Explain.

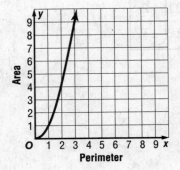

9-8 Skills Practice

Quadratic Functions

Graph each function.

1. $y = 5x^2$

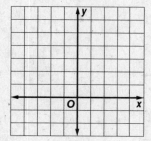

2. $y = -x^2$

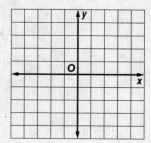

3. $y = -5x^2$

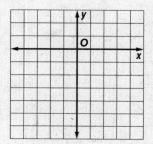

4. $y = x^2 - 1$

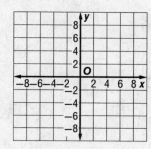

5. $y = x^2 + 4$

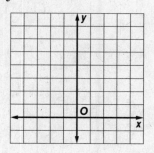

6. $y = -2x^2 + 2$

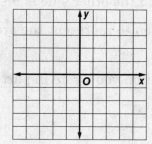

7. $y = x^2 - 4$

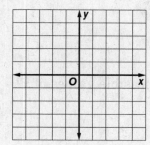

8. $y = 2x^2 - 2$

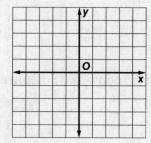

9-8　Practice

Quadratic Functions

Graph each function.

1. $y = 0.4x^2$

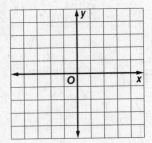

2. $y = -\dfrac{1}{2}x^2$

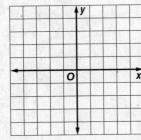

3. $y = -2x^2 - 1$

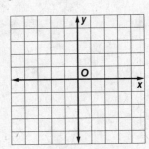

4. $y = 3x^2 - 4$

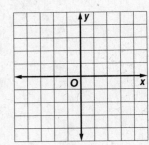

5. WINDOWS A window maker has 25 feet of wire to frame a window. One side of the window is x feet and the other side is $9 - x$ feet.

a. Write an equation to represent the area A of the window.

b. Graph the equation you wrote in part **a.**

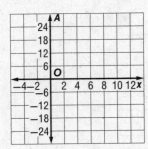

c. If the area of the window is 18 square feet, what are the two possible values of x?

9-9 Skills Practice

Cubic and Exponential Functions

Graph each function.

1. $y = 5x^3$

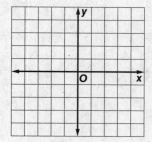

2. $y = -5x^3$

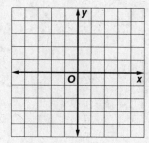

3. $y = x^3 + 4$

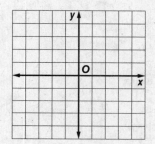

4. $y = x^3 - 4$

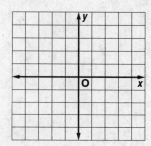

5. $y = 2x^3 + 3$

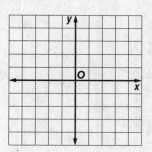

6. $y = -x^3$

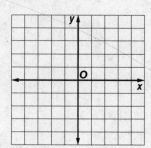

7. $y = 2^x - 4$

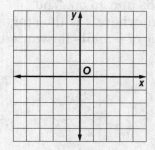

8. $y = 3^x - 7$

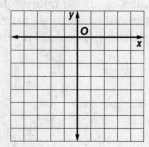

9-9 Practice

Cubic and Exponential Functions

Graph each function.

1. $y = 0.4x^3$

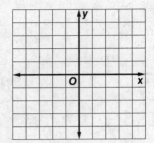

2. $y = -2x^3 - 1$

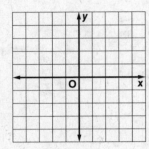

3. $y = x^3 + 0.5$

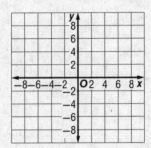

4. $y = \frac{1}{5}x^3$

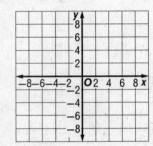

5. $y = 3^x + 0.75$

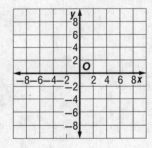

6. $y = 3^x - 4$

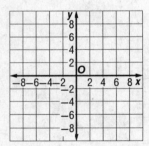

7. E–MAIL Mike forwarded an e-mail to 5 friends. Each of those 5 friends forwarded it to 5 of their friends. Each of those friends forwarded it to five friends and so on. The function $N = 5^x$ represents the total number of e-mails forwarded, where x is the stage of the e-mails. Graph the function. In what stage will the number of e-mails forwarded be at least 625?

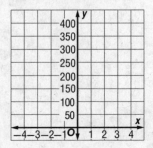

10-1 Skills Practice

Squares and Square Roots

Find each square root.

1. $\sqrt{1}$　　　　　　　　2. $\sqrt{9}$　　　　　　　　3. $\sqrt{25}$

4. $\sqrt{49}$　　　　　　　5. $\sqrt{64}$　　　　　　　6. $\sqrt{169}$

7. $-\sqrt{36}$　　　　　　8. $\sqrt{-81}$　　　　　　9. $-\sqrt{64}$

10. $-\sqrt{169}$　　　　　11. $\sqrt{-196}$　　　　　12. $-\sqrt{121}$

13. $\sqrt{225}$　　　　　14. $\sqrt{441}$　　　　　15. $\sqrt{625}$

16. $\pm\sqrt{289}$　　　　17. $\pm\sqrt{324}$　　　　18. $\pm\sqrt{8100}$

19. $\sqrt{2.25}$　　　　　20. $\sqrt{0.16}$　　　　　21. $\sqrt{3.24}$

Use a calculator to find each square root to the nearest tenth.

22. $\sqrt{31}$　　　　　　23. $\sqrt{40}$　　　　　　24. $\sqrt{94}$

25. $\sqrt{132}$　　　　　26. $-\sqrt{68}$　　　　　27. $-\sqrt{247}$

28. $\sqrt{-521}$　　　　29. $-\sqrt{314}$　　　　30. $-\sqrt{902}$

31. $-\sqrt{0.85}$　　　　32. $-\sqrt{2.45}$　　　　33. $-\sqrt{4.05}$

Estimate each square root to the nearest integer. Do not use a calculator.

34. $\sqrt{38}$　　　　　　35. $\sqrt{84}$　　　　　　36. $\sqrt{389}$

37. $\sqrt{5}$　　　　　　38. $\sqrt{118}$　　　　　39. $\sqrt{230}$

40. $-\sqrt{83}$　　　　　41. $-\sqrt{19}$　　　　　42. $-\sqrt{119}$

43. $\sqrt{9.3}$　　　　　44. $\sqrt{27.5}$　　　　45. $\sqrt{78.1}$

10-1 Practice

Squares and Square Roots

Find each square root.

1. $\sqrt{100}$ 2. $\sqrt{144}$ 3. $\sqrt{-36}$

4. $\sqrt{121}$ 5. $\sqrt{-148}$ 6. $-\sqrt{4}$

7. $-\sqrt{9}$ 8. $-\sqrt{49}$ 9. $\sqrt{256}$

10. $\sqrt{529}$ 11. $\sqrt{361}$ 12. $-\sqrt{196}$

Use a calculator to find each square root to the nearest tenth.

13. $-\sqrt{2.25}$ 14. $\sqrt{38}$ 15. $\sqrt{249}$

16. $\sqrt{131}$ 17. $\sqrt{7}$ 18. $\sqrt{52}$

19. $\sqrt{168}$ 20. $\sqrt{499}$ 21. $-\sqrt{217}$

22. $\pm\sqrt{218}$ 23. $\pm\sqrt{42}$ 24. $\pm\sqrt{94}$

25. $\pm\sqrt{50}$ 26. $\pm\sqrt{11.7}$ 27. $\pm\sqrt{208}$

28. Find the negative square root of 840 to the nearest tenth.

29. If $x^2 = 476$, what is the value of x to the nearest tenth?

30. The number $\sqrt{22}$ lies between which two consecutive whole numbers? Do not use a calculator.

Estimate each square root to the nearest integer. Do not use a calculator.

31. $\sqrt{76}$ 32. $\sqrt{123}$ 33. $\sqrt{300}$

34. $\sqrt{90}$ 35. $\sqrt{19}$ 36. $\sqrt{248}$

37. **GEOMETRY** A square tarpaulin covering a softball field has an area of 441 m². What is the length of one side of the tarpaulin?

38. **MONUMENTS** Use the equation $d = 1.22 \cdot \sqrt{h}$ where d is the distance to the horizon in miles and h is the person's distance from the ground in feet. The highest observation deck on the Eiffel Tower in Paris is about 899 feet above the ground. About how far could a visitor see on a clear day?

10-2 Skills Practice

The Real Number System

Name all of the sets of numbers to which each real number belongs. Let
W = whole numbers, Z = integers, Q = rational numbers, and I = irrational
numbers.

1. 12

2. 25

3. −5

4. $\frac{1}{8}$

5. $\frac{1}{9}$

6. 0.343434. . .

7. $\sqrt{31}$

8. $\sqrt{7}$

9. $\frac{25}{3}$

10. $-\frac{32}{4}$

11. 6.54

12. 24.6

13. 418

14. 0

15. 0.050050005 . . .

Determine whether each statement is *sometimes*, *always*, or *never* true.

16. A whole number is a rational number.

17. A rational number is a whole number.

18. A negative number is an integer.

19. Zero is an irrational number.

Replace each ● with <, >, or = to make a true statement.

20. $\sqrt{4}$ ● $2\frac{3}{7}$

21. $\sqrt{5}$ ● 2.1

22. $-\sqrt{12}$ ● −3.5

23. $\sqrt{104.04}$ ● 10.2

24. 7.8 ● $\sqrt{55}$

25. 15.1 ● $\sqrt{231}$

Order each set of numbers from least to greatest.

26. $5\frac{1}{3}$, 5.3, $\sqrt{28}$, $2\frac{1}{4}$

27. $\sqrt{53}$, $7\frac{1}{4}$, $\frac{36}{5}$, 7.27

28. −9.35, $-\sqrt{72.75}$, $-9\frac{2}{10}$, −9

ALGEBRA Solve each equation. Round to the nearest tenth, if necessary.

29. $a^2 = 64$

30. $d^2 = 169$

31. $f^2 = 441$

32. $76 = g^2$

33. $115 = h^2$

34. $k^2 = 450$

35. $b^2 = 4.41$

36. $y^2 = 0.36$

37. $m^2 = 0.0025$

135

10-2 Practice

The Real Number System

Name all of the sets of numbers to which each real number belongs. Let
W = whole numbers, Z = integers, Q = rational numbers, and I = irrational
numbers.

1. 15

2. − 41

3. $\frac{1}{4}$

4. $\frac{1}{3}$

5. 0.212121. . .

6. $\sqrt{8}$

7. $\sqrt{45}$

8. $\frac{36}{9}$

9. $-\frac{28}{7}$

10. 2.31

11. 45.6

12. 0.090090009. . .

Determine whether each statement is *sometimes*, *always*, or *never* true.

13. A decimal number is an irrational number.

14. An integer is a whole number.

15. A whole number is an integer.

16. A negative integer is a whole number.

Replace each ● with < , >, or = to make a true statement.

17. $3.2 ● \sqrt{9.5}$

18. $1\frac{1}{2} ● \sqrt{3}$

19. $\sqrt{17} ● 4.1$

20. $\sqrt{7.84} ● 2.8$

21. $1\frac{3}{4} ● \sqrt{3.0625}$

22. $3.67 ● \sqrt{12}$

Order each set of numbers from least to greatest.

23. $\sqrt{49}, 6.\overline{91}, 7\frac{1}{8}, \frac{15}{2}$

24. $4\frac{1}{3}, \sqrt{43}, \frac{12}{3}, 4.13$

25. $-2, -1.5, -1\frac{8}{10}, -\sqrt{6}$

ALGEBRA Solve each equation. Round to the nearest tenth, if necessary.

26. $h^2 = 361$

27. $k^2 = 10.24$

28. $c^2 = 111$

29. $330 = t^2$

30. $0.089 = u^2$

31. $w^2 = 0.0144$

32. **GARDENING** Ray planted a square garden that covers an area of 200 ft^2.
How many feet of fencing does he need to surround the garden?

10-3 Skills Practice

Triangles

Classify each angle as *acute*, *obtuse*, *right*, or *straight*.

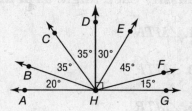

1. ∠AHB

2. ∠AHC

3. ∠AHD

4. ∠AHE

5. ∠AHF

6. ∠AHG

Find the value of *x* in each triangle. Then classify each triangle by its angles and by its sides.

7.

8.

9.

10.

11.

12.

Classify each dashed triangle by its angles and by its sides.

13.

14.

15.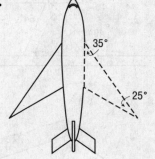

10-3 Practice

Triangles

Classify each angle as *acute, obtuse, right,* or *straight*.

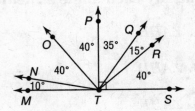

1. ∠MTN

2. ∠MTO

3. ∠MTP

4. ∠MTQ

5. ∠MTR

6. ∠NTO

Find the value of *x* in each triangle. Then classify each triangle by its angles and by its sides.

7.

8.

9.

10.

11. **ALGEBRA** The measures of the angles of a triangle are in the ratio 5:6:9. What is the measure of each angle?

12. **ALGEBRA** Determine the measures of the angles of △MNO if the measures of the angles are in the ratio 2:4:6.

Classify each triangle by its angles and by its sides.

13.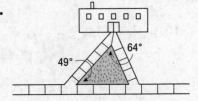

14.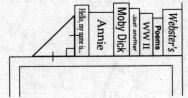

15.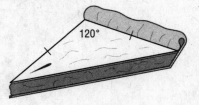

16.

10-4 Skills Practice

The Pythagorean Theorem

Find the length of the hypotenuse of each right triangle. Round to the nearest tenth, if necessary.

1.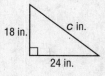

2.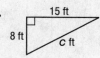

3.

4.

5.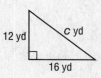

6.

If c is the measure of the hypotenuse, find each missing measure. Round to the nearest tenth, if necessary.

7. $a = ?, b = 24, c = 26$

8. $a = 16, b = ?, c = 34$

9. $a = 24, b = ?, c = 40$

10. $a = 5, b = ?, c = 7$

11. $a = ?, b = 32, c = 39$

12. $a = 21, b = ?, c = 48$

13. $a = 18, b = 29, c = ?$

14. $a = ?, b = 36, c = 49$

15. $a = 8, b = ?, c = 12$

16. $a = 14, b = 21, c = ?$

17. $a = ?, b = 30, c = 40$

18. $a = 4, b = ?, c = 7$

19. $a = 13, b = 18, c = ?$

20. $a = ?, b = 55, c = 75$

The lengths of three sides of a triangle are given. Determine whether each triangle is a right triangle.

21. 14 m, 5 m, 4 m

22. 3 in., 4 in., 5 in.

10-4 Practice

The Pythagorean Theorem

Find the length of the hypotenuse of each right triangle. Round to the nearest tenth, if necessary.

1.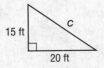

2.

3.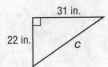

4.

5.

6.

If c is the measure of the hypotenuse, find each missing measure. Round to the nearest tenth, if necessary.

7. $a = ?, b = 15, c = 31$

8. $a = 8, b = ?, c = 16$

9. $a = 11, b = 16, c = ?$

10. $a = ?, b = 13, c = 19$

11. $a = 10, b = ?, c = 18$

12. $a = 21, b = 23, c = ?$

13. $a = ?, b = 27, c = 35$

14. $a = 48, b = ?, c = 61$

15. $a = 26, b = \sqrt{596}, c = ?$

16. $a = ?, b = 12, c = \sqrt{318}$

The lengths of three sides of a triangle are given. Determine whether each triangle is a right triangle.

17. 5 m, 5 m, 10 m

18. 9 in., 12 in., 15 in.

19. ARCHITECTURE The diagonal distance covered by a flight of stairs is 21 ft. If the stairs cover 10 ft horizontally, how tall are they?

20. KITES A kite is flying at the end of a 300-foot string. It is 120 feet above the ground. About how far away horizontally is the person holding the string from the kite?

10-5 Skills Practice

The Distance Formula

Find the distance between each pair of points. Round to the nearest tenth, if necessary.

1. $A(2, 4)$, $B(1, 3)$

2. $P(5, 10)$, $Q(-1, 1)$

3. $G(3, -1)$, $H(5, 6)$

4. $C(-2, -6)$, $D(-7, 1)$

5. $E(-6, 2)$, $F(4, 1)$

6. $J(-5, -3)$, $K(4, -2)$

7. $M(-5, -5)$, $N(3, -4)$

8. $V(4, 7)$, $W(1, 6)$

9. $X(4, 6)$, $Y(-3, -7)$

10. $R(0, 0)$, $S(-1, -1)$

11. $T(7, 3)$, $U(-2, -2)$

12. $A(6, 2)$, $B(1, 3)$

13. $V(2, -6)$, $W(4, -7)$

14. $C(6, 2)$, $D(4, 7)$

15. $X(7, 8)$, $Y(-7, 1)$

16. $E(7, 3)$, $F(-1, 4)$

17. $A(5, 10)$, $B(-4, -3)$

18. $G(-6, 2)$, $H(2, 4)$

GEOMETRY Classify each triangle by its sides. Then find the perimeter of each triangle. Round to the nearest tenth.

19.
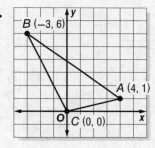

20.
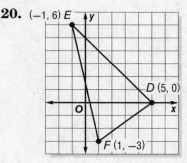

GEOMETRY The coordinates of the vertices of a triangle are given. Find the perimeter of each triangle. Round to the nearest tenth, if necessary.

21. $J(4, 5)$, $K(-2, 2)$, and $L(-4, 4)$

22. $E(3, 5)$, $F(4, 8)$, and $G(-1, 6)$

23. $X(8, 1)$, $Y(3, 3)$, and $Z(5, -3)$

24. $A(-3, 5)$, $B(-3, -1)$, and $C(7, -1)$

10-5 Practice

The Distance Formula

Find the distance between each pair of points. Round to the nearest tenth, if necessary.

1. $A(5, 2), B(3, 4)$

2. $C(-2, -4), D(1, 3)$

3. $E(-3, 4), F(-2, 1)$

4. $G(0, 0), H(-7, 8)$

5. $R(-4, -8), S(2, -3)$

6. $G(9, 9), H(-9, -9)$

7. $M(1, 1), N(-10, -10)$

8. $P\left(1\frac{1}{2}, 3\right), Q\left(5, 6\frac{1}{4}\right)$

9. $R\left(7, 4\frac{1}{2}\right), S\left(6\frac{1}{2}, 3\frac{1}{4}\right)$

10. $T\left(-3\frac{1}{2}, -4\frac{1}{4}\right), U\left(5\frac{1}{2}, 1\frac{1}{2}\right)$

11. $A(5, 1), B(-4, 23)$

12. $V(4, 6), W(-8, -12)$

13. $C(-2, -4), D(-5, 6)$

14. $X(1, -7), Y(-1, 7)$

15. $E(5, -3), F(-7, 8)$

16. $A(8, 8), B(-8, -8)$

GEOMETRY Classify each triangle by its sides. Then find the perimeter of each triangle. Round to the nearest tenth.

17.

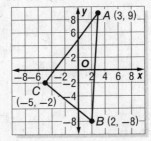

18.

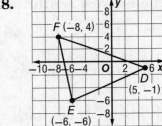

19. MAPS On a map of the school, the baseball field is located at the coordinates (1, 7). The front entrance of the school is located at (5, 2). If each coordinate unit corresponds to 10 yards, how far is it from the front entrance to the baseball field?

20. Determine whether $\triangle XYZ$ with vertices $X(3, 4), Y(2, -3)$, and $Z(-5, -2)$ is isosceles. Explain your answer.

21. Is $\triangle DEF$ with vertices $D(1, 4), E(6, 2), F(-1, 3)$ a scalene triangle? Explain.

10-6 Skill Practice

Special Right Triangles

Find each missing measure.

1.
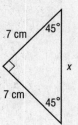
.7 cm 45°
x
7 cm 45°

2.
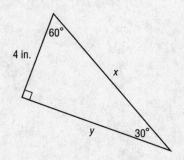
60°
4 in.
x
y 30°

3.
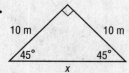
10 m 10 m
45° 45°
x

4.
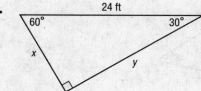
24 ft
60° 30°
x
y

5.
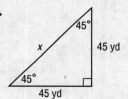
45°
x 45 yd
45°
45 yd

6.
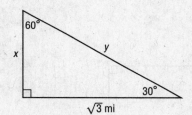
60°
x y
30°
√3 mi

7.
x
60°
y 14 mm
30°

8.
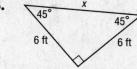
x
45° 45°
6 ft 6 ft

9.

25 in. 45°
x
25 in. 45°

10.
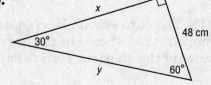
x
48 cm
30°
y 60°

10-6 Practice

Special Right Triangles

Find each missing measure.

1.
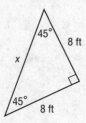
45°
8 ft
x
45°
8 ft

2.
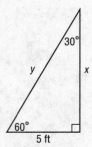
30°
y
x
60°
5 ft

3.
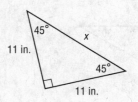
45°
x
11 in.
45°
11 in.

4.
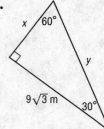
x
60°
y
$9\sqrt{3}$ m
30°

5.
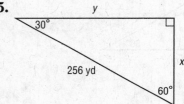
y
30°
256 yd
x
60°

6.
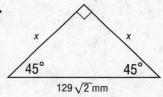
x
x
45°
45°
$129\sqrt{2}$ mm

7.
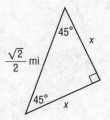
45°
x
$\frac{\sqrt{2}}{2}$ mi
45°
x

8.
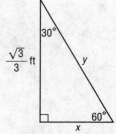
30°
$\frac{\sqrt{3}}{3}$ ft
y
60°
x

9. **SHORTCUTS** To get to school, Hari takes a shortcut across a square-shaped lot as shown in the drawing at the right. What is the distance of the shortcut Hari takes?

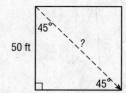
45°
?
50 ft
45°

10. **LADDERS** A ladder leaning against the side of a building forms a 60° angle with the ground. If the ladder is 20 feet long, how far from the building is the base of the ladder?

11-1 Skills Practice

Angle and Line Relationships

In the figure at the right, $c \parallel d$ and p is a transversal.
If $m\angle 5 = 110°$, find the measure of each angle.

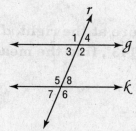

1. $\angle 6$ 2. $\angle 8$

3. $\angle 2$ 4. $\angle 4$

In the figure at the right, $g \parallel k$ and r is a transversal.
If $m\angle 7 = 60°$, find the measure of each angle.

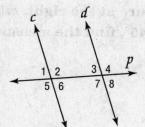

5. $\angle 4$ 6. $\angle 6$

7. $\angle 5$ 8. $\angle 3$

Classify the pairs of angles shown. Then find the value of x in each figure.

9.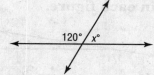
120° x°

10.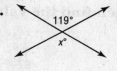
119°
x°

11.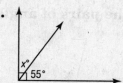
x°
55°

12.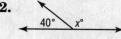
40° x°

13.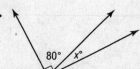
80° x°

14.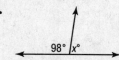
98° x°

15.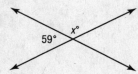
22°
x°

16.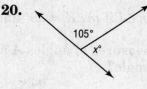
59° x°

17.
x°   6°

18.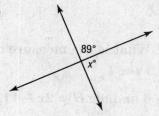
89°
x°

19.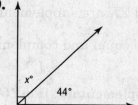
x°
44°

20.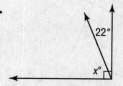
105°
x°

11-1 **Practice**

Angle and Line Relationships

In the figure at the right, $m \parallel n$ and r is a transversal.
If $m\angle 2 = 45°$, find the measure of each angle.

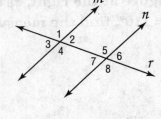

1. $\angle 4$ 2. $\angle 5$

3. $\angle 7$ 4. $\angle 8$

5. $\angle 6$ 6. $\angle 3$

In the figure at the right, $d \parallel e$ and a is a transversal.
If $m\angle 5 = 143°$, find the measure of each angle.

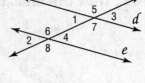

7. $\angle 7$ 8. $\angle 6$

9. $\angle 4$ 10. $\angle 2$

11. $\angle 1$ 12. $\angle 8$

Classify the pairs of angles shown. Then find the value of x in each figure.

13.

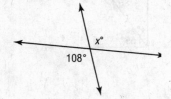

14.

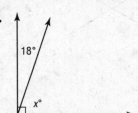

15.

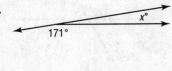

16.

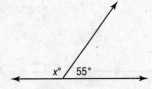

17.

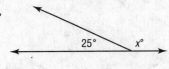

18.

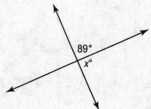

19. Angles Q and R are complementary. Find $m\angle R$ if $m \angle Q = 24°$.

20. Find $m\angle J$ if $m\angle K = 29°$ and $\angle J$ and $\angle K$ are supplementary.

21. The measures of angles A and B are equal and complementary. What is the measure of each angle?

22. **ALGEBRA** Angles G and H are complementary. If $m\angle G = 3x + 6$ and $m\angle H = 2x - 11$, what is the measure of each angle?

146

11-2 Skills Practice

Congruent Triangles

Name the corresponding parts in each pair of congruent triangles. Then complete the congruence statement.

1.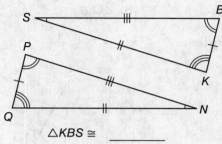

△KBS ≅ _____

2.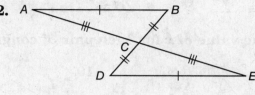

△ACB ≅ _____

Complete each congruence statement if △MRU ≅ △ACF.

3. ∠R ≅ __?__   4. $\overline{CA}$ ≅ __?__   5. $\overline{MU}$ ≅ __?__   6. ∠A ≅ __?__

Find the value of x for each pair of congruent triangles.

7.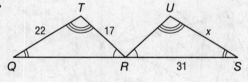

8.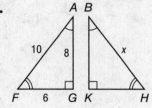

9.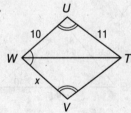

Determine whether the triangles shown are congruent. If so, name the corresponding parts and write a congruence statement.

10.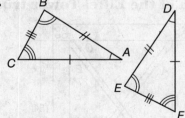

11.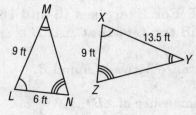

11-2 Practice

Congruent Triangles

Complete the congruence statement if △CMH ≅ △PLF and △DNO ≅ △AET.

1. ∠M ≅ __?__   2. $\overline{MC}$ ≅ __?__   3. $\overline{DN}$ ≅ __?__   4. ∠A ≅ __?__

5. $\overline{FL}$ ≅ __?__   6. ∠C ≅ __?__   7. $\overline{TE}$ ≅ __?__   8. ∠O ≅ __?__

Find the value of x for each pair of congruent triangles.

9.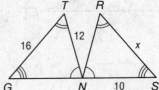

10.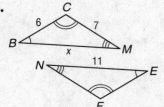

11.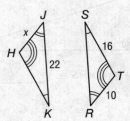

12. ALGEBRA If △DEC ≅ △PRM, what is the value of x?

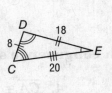

Determine whether the triangles shown are congruent. If so, name the corresponding parts and write a congruence statement.

13.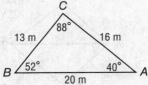

14.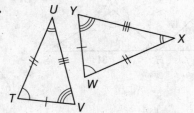

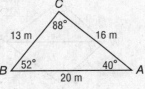

ARCHITECTURE For Exercises 15 and 16, use the diagram of the Eiffel Tower truss at the right and the fact that △ACB ≅ △DFE.

15. Find the distance between A and B.

16. What is the measure of ∠B?

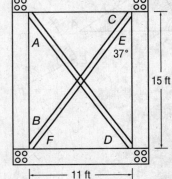

148

11-3 Skills Practice

Rotations

Draw each figure after the rotation described.

1. 90° clockwise rotation
about point *B*

2. 180° clockwise rotation
about point *C*

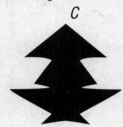

3. A figure has vertices *A*(1, 1), *B*(1, 3),
C(3, 3), and *D*(4, 1). Graph the figure
and its image after a rotation
of 90° clockwise about the origin.

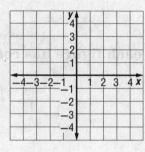

Determine whether each figure has rotational symmetry. If it does, describe the angle of rotation.

4.

5.

6.

7.

11-3 Practice

Rotations

Draw each figure after the rotation described.

1. 270° clockwise rotation about point *A*

2. 180° clockwise rotation about point *A*

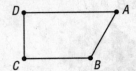

3. A figure has vertices *A*(1, 3), *B*(1, 5), and *C*(5, 4). Graph the figure and its image after a rotation of 90° clockwise about the origin.

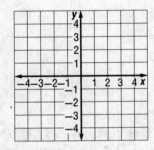

Determine whether each figure has rotational symmetry. If it does, describe the angle of rotation.

4.

5.

6. FLAGS Many countries have cooperated to build the International Space Station. The flags below represent three of them.

United Kingdom

Switzerland

Sweden

a. Which flags have rotational symmetry?

b. Describe the angle of rotation for each flag.

11-4 Skills Practice

Quadrilaterals

Find the value of *x* in each quadrilateral. Then find the missing angle measures.

1.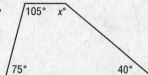

2.

3.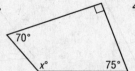

4.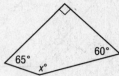

5.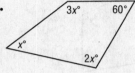

6.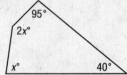

7.

8.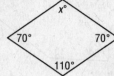

9.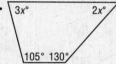

10.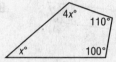

11.

12.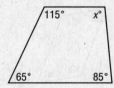

Classify each quadrilateral using the name that *best* describes it.

13.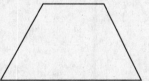

14.

15.

16.

17.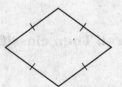

18.

Tell whether each statement is *sometimes, always,* or *never* true.

19. A rhombus is a square.

20. A square is a parallelogram.

21. A parallelogram is a square.

11-4 Practice

Quadrilaterals

Find the value of x in each quadrilateral. Then find the missing angle measures.

1.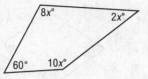

2.

3.

4.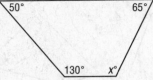

5.

6.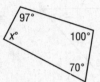

7.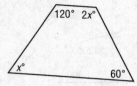

8.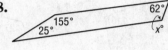

9.

Tell whether each statement is *sometimes*, *always*, or *never* true.

10. A parallelogram is a trapezoid.

11. A square is a quadrilateral.

12. A rhombus is a rectangle.

13. A quadrilateral is a rectangle.

Make a drawing of each quadrilateral. Then classify each quadrilateral using the name that *best* describes it.

14. In quadrilateral $ACFG$, $m\angle A = 60°$, $m\angle C = 120°$, $m\angle F = 115°$, and $m\angle G = 65°$.

15. In quadrilateral $EMNP$, $m\angle E = 90°$, $m\angle M = 80°$, $m\angle N = 60°$, and $m\angle P = 130°$.

11-5 Skills Practice

Polygons

Determine whether the figure is a polygon. If it is, classify the polygon. If it is not a polygon, explain why.

1.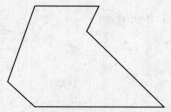

2.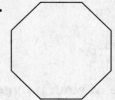

3.

4.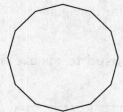

5.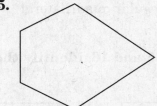

6.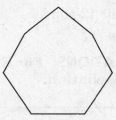

Find the sum of the measures of the interior angles of each polygon.

7. pentagon 　　　 8. 20-gon 　　　 9. nonagon 　　　 10. decagon

Find the measure of an interior angle of each polygon.

11. regular hexagon 　　　 12. regular heptagon 　　　 13. regular quadrilateral

14. regular octagon 　　　 15. regular pentagon 　　　 16. regular 100-gon

TESSELLATIONS For Exercises 17 and 18, identify the polygons used to create each tessellation.

17.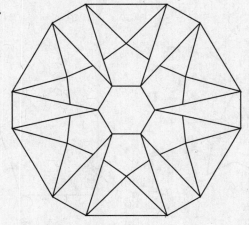

18.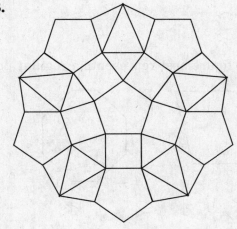

11-5 Practice

Polygons

Find the sum of the measures of the interior angles of each polygon.

1. quadrilateral

2. decagon

3. 12-gon

4. heptagon

5. pentagon

6. hexagon

7. 25-gon

8. 100-gon

Find the measure of an interior angle of each polygon.

9. regular nonagon

10. regular octagon

11. regular hexagon

12. regular 12-gon

13. regular quadrilateral

14. regular decagon

TESSELLATIONS For Exercises 15 and 16, identify the polygons used to create each tessellation.

15.
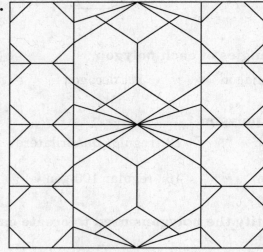

16.
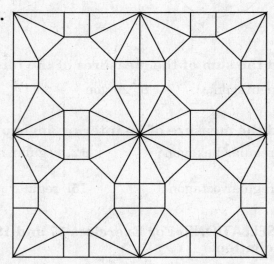

17. Which figure best represents a regular polygon?

A
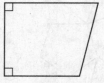

B

C

D
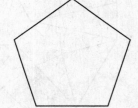

11-6 Skills Practice

Area of Parallelograms, Triangles, and Trapezoids

Find the area of each figure.

1.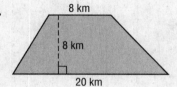
9 yd 10 yd
4.5 yd

2.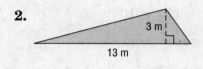
3 m
13 m

3.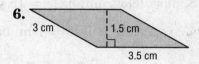
9 km
6 km
2 km

4.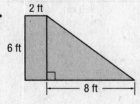
8 km
8 km
20 km

5.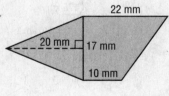
4 m
11 m

6.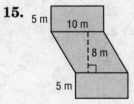
3 cm
1.5 cm
3.5 cm

Find the area of each figure.

7. triangle: base = 11 m; height = 3 m

8. parallelogram: base = 8 cm; height = 9.5 cm

9. trapezoid: height = 12 yd; bases = 4 yd, 7 yd

10. parallelogram: base = 6.5 ft; height = 12 ft

11. trapezoid: height = 10 m; bases = 3 m, 6 m

12. triangle: base = 7 km; height = 5 km

Find the area of each figure.

13.
2 ft
6 ft
8 ft

14.
22 mm
20 mm 17 mm
10 mm

15.
5 m
10 m
8 m
5 m

GEOGRAPHY For Exercises 16–18, use the approximate measurements to estimate the area of each state.

16. Alabama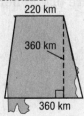
220 km
360 km
360 km

17. Florida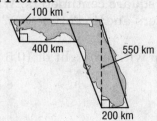
100 km
400 km 550 km
200 km

18. Nevada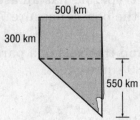
500 km
300 km
550 km

11-6 Practice

Area of Parallelograms, Triangles, and Trapezoids

Find the area of each figure.

1. parallelogram: base = 12 m; height = 10 m

2. trapezoid: height = 13 cm; bases = 3 cm, 7 cm

3. triangle: base = 9.4 ft; height = 5 ft

4. triangle: base = 8.5 km; height = 14 km

5. parallelogram: base = 15 yd; height = 7 yd

6. trapezoid: height = 7 m; bases = 6 m, 9 m

Find the area of each figure.

7.
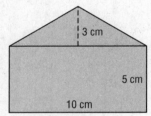

8.
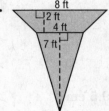

9.
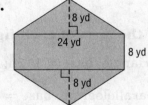

GEOGRAPHY For Exercises 10–12, use the approximate measurements to estimate the area of each state.

10. Maine
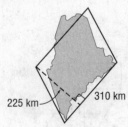

11. Idaho
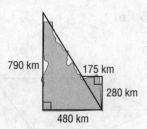

12. North Carolina
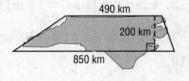

13. Suppose a triangle has an area of 220 square meters. What is the measure of the height if the base measures 20 meters?

14. A trapezoid has an area of 27.5 square centimeters. What is the measure of the height if the bases measure 7 centimeters and 4 centimeters?

15. Find the base of a parallelogram with a height of 10.5 feet and an area of 189 square feet.

11-7 Skills Practice

Circles and Circumference

Find the circumference of each circle. Round to the nearest tenth.

1. 9 m

2. 17 ft

3. 3 yd

4. 5 cm

5. The radius is 7 kilometers.

6. The diameter is 20 centimeters.

7. The diameter is 8.5 meters.

8. The radius is 11 yards.

9. The diameter is $6\frac{2}{5}$ feet.

10. The radius is 25 inches.

Match each circle described in the column on the left with its corresponding measurement in the column on the right.

11. diameter: 6 units   a. circumference: 18.8 units

12. radius: 9 units   b. circumference: 40.8 units

13. diameter: 13 units   c. circumference: 15.7 units

14. radius: 2.5 units   d. circumference: 56.5 units

15. **SPORTS** A basketball goal is 18 inches in diameter. A basketball has a diameter of about 9.6 inches. What is the difference in circumference between the goal and the center cross-section of a basketball?

16. **CULTURE** The Navajo and Pueblo Indians create large, circular sand paintings as part of traditional healing ceremonies. How much more circumference does a sand painting with a 20-foot diameter have compared with one with a 5-foot diameter?

17. **SPORT** In bowling, the distance from the foul line to the headpin is 60 feet. A bowling ball has a radius of about 4.3 inches. How many times must the ball rotate in order to strike the headpin?

Find the perimeter of each figure. Round to the nearest tenth.

18.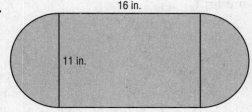
16 in.
11 in.

19.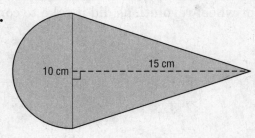
10 cm
15 cm

11-7 Practice

Circles and Circumference

Find the circumference of each circle. Round to the nearest tenth.

1. The diameter is 18 yards.

2. The radius is 4 meters.

3. The diameter is 4.2 meters.

4. The radius is 4.5 feet.

5. The radius is $9\frac{3}{4}$ miles.

6. The diameter is 6 kilometers.

7. The diameter is $2\frac{5}{8}$ inches.

8. The radius is $11\frac{3}{16}$ centimeters.

Match each circle described in the column on the left with its corresponding measurement in the column on the right.

9. radius: 8.5 units

a. circumference: 53.4 units

10. diameter: 9 units

b. circumference: 20.4 units

11. diameter: 6.5 units

c. circumference: 28.3 units

12. radius: 12 units

d. circumference: 75.4 units

13. SPORTS A baseball has a radius of about 1.5 inches. Home plate is 16 inches wide. If a baseball were rolled across home plate, how many complete rotations would it take to cover the distance?

14. SPORTS A soccer ball has a circumference of about 28 inches, while the goal is 24 feet wide. How many soccer balls would be needed to cover the distance between the goalposts?

15. HISTORY Chariot races reached their peak in popularity in ancient Rome around the 1st and 2nd centuries A.D. A chariot wheel had a radius of about one foot. One lap around the track in the Circus Maximus was approximately 2,300 feet. How many chariot-wheel revolutions did it take to complete one lap?

11-8 Skills Practice

Area of Circles

Find the area of each circle. Round to the nearest tenth.

1.
5 cm

2.
24 m

3.
20 ft

4.
8 in.

5.
6 yd

6.
18 mi

7. radius = 6 kilometers

8. diameter = 14 inches

9. diameter = 6 yards

10. radius = 5 feet

11. radius = 18 centimeters

12. diameter = 8.4 meters

Find the area of each shaded sector. Round to the nearest tenth.

13.
5 yd

14.
9 m
40°

15.
300°
10 ft

16.
72° 7 in.

17. radius = 4 centimeters;
central angle = 12°

18. radius = 11 miles;
central angle = 270°

19. radius = 3 feet;
central angle = 60°

20. radius = 15 meters;
central angle = 120°

11-8 Practice

Area of Circles

Find the area of each circle. Round to the nearest tenth.

1.
5 cm

2.
17 m

3.
9.2 in.

4.
11.5 yd

5. diameter = 9 kilometers

6. radius = 21 inches

7. diameter = 19.8 yards

8. radius = 7.3 feet

9. radius = 0.5 centimeter

10. diameter = 6.4 meter

Find the area of each shaded sector. Round to the nearest tenth.

11.
108°
2.8 in.

12.
25°
30 ft

13.
67°
12.6 m

14.
204°
100 cm

15. radius = 0.75 mile; central angle: 86°

16. radius = 33.3 kilometers; central angle: 349°

17. **CONTAINERS** The top of a soda can has a diameter of 6 cm. What is the area of the top of the can to the nearest tenth of a centimeter?

18. **COOKIES** What is the difference in area between a cookie cut from a cutter that has a diameter of 4 inches and a cookie cut from a cutter with a radius of 3 inches?

19. **PIZZA** Pizza Palace's largest pizza box has side lengths of 18 inches. A customer wants to special order a pizza with an area of 300 square inches. Will the pizza fit in one of Pizza Palace's pizza boxes? Explain.

11-9 Skills Practice

Area of Composite Figures

Find the area of each figure. Round to the nearest tenth, if necessary.

1.
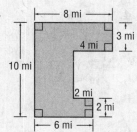

2.
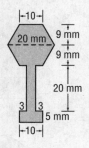

3.
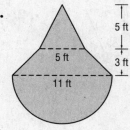

4.
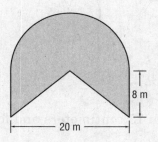

5.
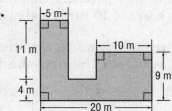

6.
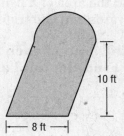

7.
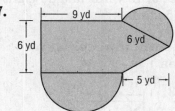

8.
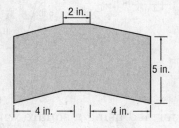

9.
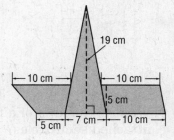

10. What is the area of a figure formed using a square with sides of 12 kilometers and three circles with diameters of 12 kilometers each?

Find the area of each shaded area. Round to the nearest tenth, if necessary. (*Hint:* Find the total area and subtract the non-shaded area.)

11.
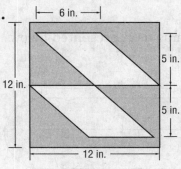

12.
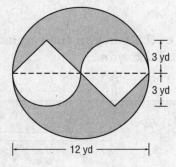

13.
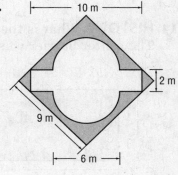

11-9 Practice

Area of Composite Figures

Find the area of each figure. Round to the nearest tenth, if necessary.

1.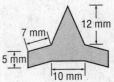

2.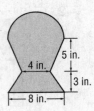

3.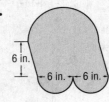

4.

5.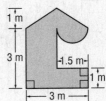

6.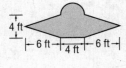

7.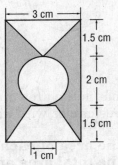

8.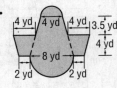

9. What is the area of a figure formed using a square with sides of 15 centimeters and four attached semicircles?

10. Find the area of a figure formed using a parallelogram with a base of 10 yards and a height of 12 yards and two triangles with bases of 10 yards and heights of 5 yards.

Find the area of each shaded area. Round to the nearest tenth, if necessary. (Hint: Find the total area and subtract the non-shaded area.)

11.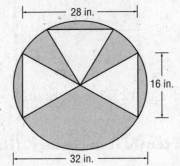

12.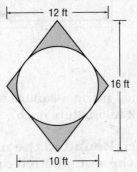

13.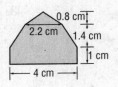

14. HISTORY What is the area of the track in the Circus Maximus as represented below? The center barrier was named the *spina*.

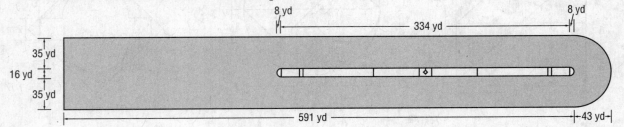

12-1 Skills Practice

Three-Dimensional Figures

Identify each figure. Name the bases, faces, edges, and vertices.

1.

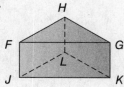

2.

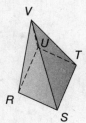

3.

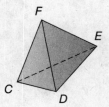

4.

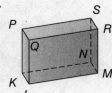

Draw and describe the shape resulting from each cross section.

5.

6.

7.

8.

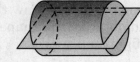

12-1 Practice

Three-Dimensional Figures

Identify each figure. Name the bases, faces, edges, and vertices.

1.

2.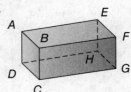

3.

Draw and describe the shape resulting from each cross section.

4.

5.

6.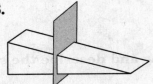

7. **GLOBES** Miguel has a globe on his shelf. Draw and describe the shape resulting from vertical, angled, and horizontal cross sections of the globe.

vertical   angled   horizontal

8. **CRYSTALS** Janie collects rocks and minerals. She bought the beryl crystal shown at the right. Draw the top view and side view. Then draw and describe the shape resulting from an angled cross section of the figure.

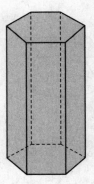

top view   side view   angled cross
                                    section

12-2 Skills Practice

Volume of Prisms

Find the volume of each figure. If necessary, round to the nearest tenth.

1.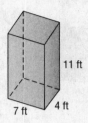
11 ft
7 ft 4 ft

2.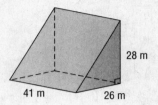
28 m
41 m 26 m

3.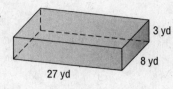
3 yd
8 yd
27 yd

4.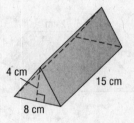
4 cm
15 cm
8 cm

5.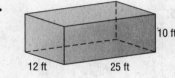
10 ft
12 ft 25 ft

6.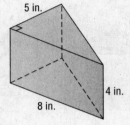
5 in.
4 in.
8 in.

7. Rectangular prism: length 18 feet, width 9 feet, height 1 foot

8. Triangular prism: base of triangle 22 yards, height of triangle 14 yards, height of prism 30 yards

9. Find the width of a rectangular prism with a length of 11.5 inches, a height of 14 inches, and a volume of 483 cubic inches.

10. BLUEPRINTS The blueprints for a barn are shown at the right. What is the volume of the barn?

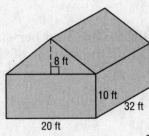
8 ft
10 ft
32 ft
20 ft

11. SCULPTURES The artist that created the sculpture shown needs to pack it for shipping. What is the volume of the sculpture?

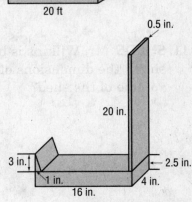
0.5 in.
20 in.
3 in.
1 in.
2.5 in.
4 in.
16 in.

12-2 Practice

Volume of Prisms

Find the volume of each figure. If necessary, round to the nearest tenth.

1.
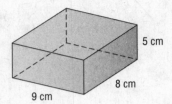
5 cm
8 cm
9 cm

2.
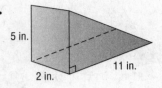
5 in.
11 in.
2 in.

3.
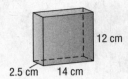
12 cm
2.5 cm   14 cm

4.
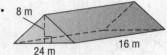
8 m
16 m
24 m

5.
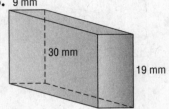
9 mm
30 mm
19 mm

6.
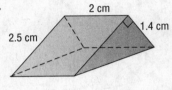
2 cm
1.4 cm
2.5 cm

7. Rectangular prism: length 22.5 feet, width 12.5 feet, height 1.2 feet

8. Triangular prism: base of triangle 17 centimeters, height of triangle 3 centimeters, height of prism 10.2 centimeters

9. Find the height of a rectangular prism with a length of 11 meters, a width of 0.5 meter, and a volume of 23.1 cubic meters.

10. **FORTS** Gina and her sister built a fort out of boxes. What is the volume of the fort?

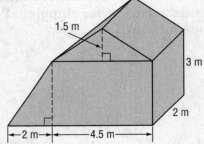
3 ft
4 ft
2 ft   1.75 ft   4 ft
2.5 ft   3 ft   2.5 ft

11. **SHEDS** Mr. Wilkins is building a shed. The sketch shows the dimensions of the shed. What is the volume of the shed?

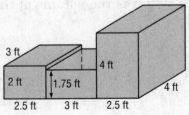
1.5 m
3 m
2 m
2 m   4.5 m

12-3 Skills Practice

Volume of Cylinders

Find the volume of each cylinder. Round to the nearest tenth.

1.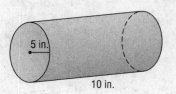
5 in.
10 in.

2.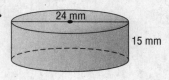
24 mm
15 mm

3.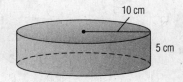
10 cm
5 cm

4. radius: 12.4 m
   height: 5.2 m

5. radius: 5.5 ft
   height: 14 ft

6. diameter: 12.5 in.
   height: 16.25 in.

Find the height of each cylinder. Round to the nearest tenth.

7. Volume: 1,494.1 m³

6.7 m

8. Volume: 1,073.9 ft³
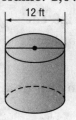
12 ft

9. Volume: 31.2 m³
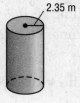
2.35 m

10. radius: 4.6 cm
    volume: 850.5 cm³

11. radius: 17 ft
    volume: 1,361.2 ft³

12. diameter: 32 yd
    volume: 3,215.4 yd³

Find the volume of each figure. Round to the nearest tenth.

13.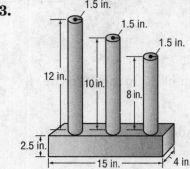
1.5 in. 1.5 in. 1.5 in.
12 in. 10 in. 8 in.
2.5 in.
15 in. 4 in.

14.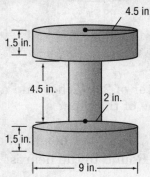
4.5 in.
1.5 in.
4.5 in.
2 in.
1.5 in.
9 in.

15.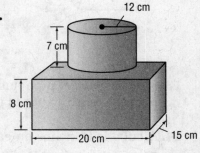
12 cm
7 cm
8 cm
20 cm
15 cm

16.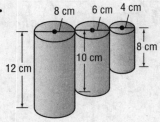
8 cm 6 cm 4 cm
8 cm
12 cm 10 cm

17.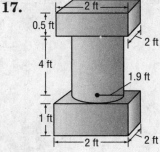
2 ft
0.5 ft
2 ft
4 ft
1.9 ft
1 ft
2 ft 2 ft

18.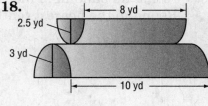
8 yd
2.5 yd
3 yd
10 yd

12-3 Practice

Volume of Cylinders

Find the volume of each cylinder. Round to the nearest tenth.

1.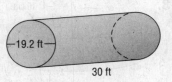

2.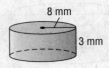

3.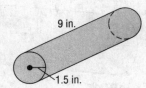

4. radius: 2.6 m
   height: 8.4 m

5. diameter: 21 ft
   height: 13 ft

6. diameter: 4.5 yd
   height: 55 yd

Find the height of each cylinder. Round to the nearest tenth.

7. radius: 15 ft
   volume: 60,052.5 ft³

8. diameter: 2.9 cm
   volume: 35.6 cm³

9. diameter: 16 in.
   volume: 5,425.9 in³

Find the volume of each figure. Round to the nearest tenth.

10.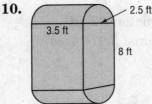

11.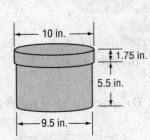

12.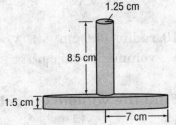

13. **PLUMBING** A pipe has a diameter of 2.5 inches and a length of 15 inches. To the nearest tenth, what is the volume of the pipe?

14. A cube is 8 inches on each side. What is the height of a cylinder having the same volume, if its radius is 4 inches? Round to the nearest tenth.

15. **FITNESS** Gary's Gym has a giant dumbbell on the roof of its building. Find the volume of the dumbbell. Round to the nearest tenth.

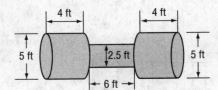

12-4 Skills Practice

Volume of Pyramids, Cones, and Spheres

Find the volume of each figure. Round to the nearest tenth, if necessary.

1.
4 ft

2.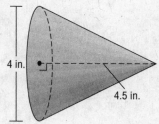
16 m
7.5 m
7.5 m

3.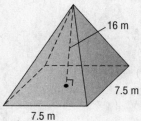
4 in.
4.5 in.

4.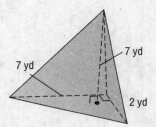
7 yd
7 yd
2 yd

5.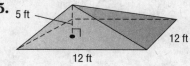
5 ft
12 ft
12 ft

6.
15 in.

7.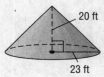
20 ft
23 ft

8.
5 mm

9.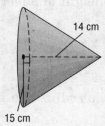
14 cm
15 cm

10. Rectangular pyramid: length 7 feet, width 2.5 feet, height 8 feet

11. Cone: radius 20 centimeters, height 30 centimeters

12. Sphere: radius 2 inches

12-4 Practice

Volume of Pyramids, Cones, and Spheres

Find the volume of each figure. Round to the nearest tenth, if necessary.

1.
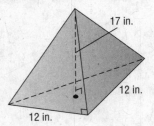
17 in.
12 in.
12 in.

2.
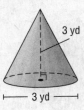
3 yd
├── 3 yd ──┤

3.

7 ft

4.
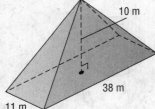
10 m
38 m
11 m

5.

4.5 cm

6.
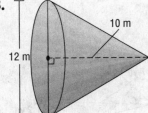
10 m
12 m

7. Find the volume of a rectangular pyramid with a length of 14 feet, a width of 12 feet, and a height of 9 feet.

8. Find the radius of a sphere with a volume of 972π cm^3.

9. Find the height of a cone with a radius of 12 in. and a volume of 408π in^3.

10. **CONTAINERS** A cone with a diameter of 3 inches has a height of 4 inches. A 2-inch square pyramid is being designed to hold nearly the same amount of ice cream. What will be the height of the square pyramid? Round to the nearest tenth.

12-5 Skills Practice

Surface Area of Prisms

Find the lateral and surface area of each prism. Round to the nearest tenth, if necessary.

1.
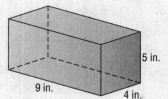
5 in.
9 in.
4 in.

2.
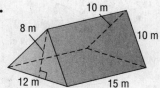
10 m
8 m
10 m
12 m
15 m

3.
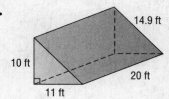
14.9 ft
10 ft
20 ft
11 ft

4.
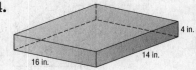
4 in.
16 in.
14 in.

5.
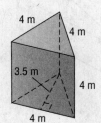
4 m
4 m
3.5 m
4 m
4 m

6.
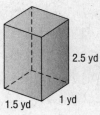
2.5 yd
1.5 yd
1 yd

7.
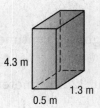
4.3 m
1.3 m
0.5 m

8.
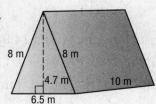
8 m
8 m
4.7 m
10 m
6.5 m

9.
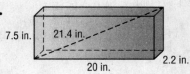
7.5 in.
21.4 in.
20 in.
2.2 in.

10. Rectangular prism: length 17 yards, width 4.5 yards, height 3 yards

11. Rectangular prism: length 16 feet, width 12 feet, height 42 feet

12. Rectangular prism: length 20.2 centimeters, width 10 centimeters, height 43 centimeters

12-5 Practice

Surface Area of Prisms

Find the lateral and surface area of each prism. Round to the nearest tenth, if necessary.

1.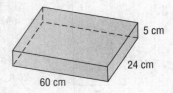
5 cm
24 cm
60 cm

2.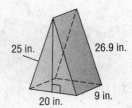
25 in.
26.9 in.
20 in.
9 in.

3.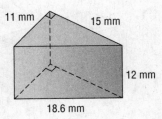
11 mm
15 mm
12 mm
18.6 mm

4.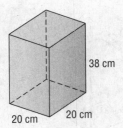
38 cm
20 cm
20 cm

5.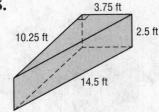
3.75 ft
10.25 ft
2.5 ft
14.5 ft

6.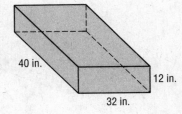
40 in.
12 in.
32 in.

7. Rectangular prism: length 10.2 meters, width 8.5 meters, height 9.1 meters

8. Rectangular prism: length 15.4 centimeters, width 14.9 centimeters, height 0.8 centimeter

9. Rectangular prism: length 28 millimeters, width 25 millimeters, height 32 millimeters

10. **DECORATING** A door that is 30 inches wide, 84 inches high, and 1.5 inches thick is to be decoratively wrapped in gift paper. How many square inches of gift paper are needed?

11. **PAINTING** The giant slice of cake on the roof of Jolene's Bake Shop needs to be repainted. Before the color can be painted on, the entire cake must be painted with primer. Calculate the surface area of the cake. Then determine how many quarts of primer Jolene needs to buy if 1 quart covers about 90 square feet.

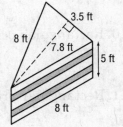
3.5 ft
8 ft
7.8 ft
5 ft
8 ft

12-6 Skills Practice

Surface Area of Cylinders

Find the lateral and surface area of each cylinder. Round to the nearest tenth.

1.
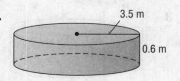
3.5 m
0.6 m

2.
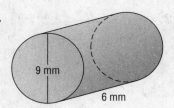
9 mm
6 mm

3.
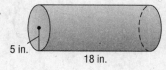
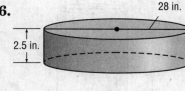
5 in.
18 in.

4.
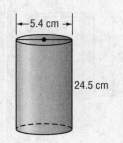
←5.4 cm→
24.5 cm

5.
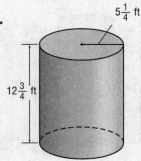
$5\frac{1}{4}$ ft
$12\frac{3}{4}$ ft

6.
28 in.
2.5 in.

7.
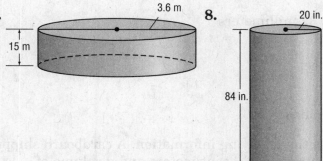
3.6 m
15 m

8.
20 in.
84 in.

9.
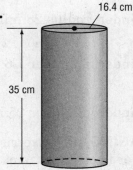
16.4 cm
35 cm

10. Cylinder: radius 16 feet, height 42 feet

11. Cylinder: diameter 20.2 centimeters, height 43 centimeters

12. Cylinder: diameter 38.2 meters, height 50 meters

12-6 Practice

Surface Area of Cylinders

Find the lateral and surface area of each cylinder. Round to the nearest tenth.

1.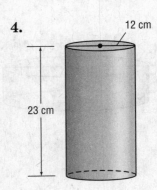
4 ft   36 ft

2.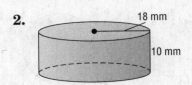
18 mm
10 mm

3.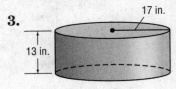
17 in.
13 in.

4.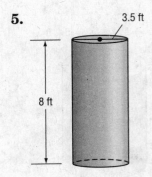
12 cm
23 cm

5.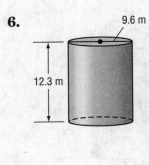
3.5 ft
8 ft

6.
9.6 m
12.3 m

7. Cylinder: radius 28 millimeters, height 32 millimeters

8. Cylinder: diameter 1.6 feet, height 4.2 feet

9. Cylinder: diameter 25 inches, height 18 inches

PACKAGING For Exercises 10 and 11, use the following information. A cardboard shipping container is in the form of a cylinder, with a radius of 6 centimeters and a volume of 8595.4 cubic centimeters.

10. Find the length of the shipping container. Round to the nearest tenth.

11. Find the surface area of the shipping container. Round to the nearest tenth.

12. **PAINTING** Mr. Jenkins has cylindrical columns and rectangular prism posts on his front porch. Both have a height of 3.5 feet. The columns have a radius of 0.5 foot. The prisms have a length and width of 1.25 feet. There are 4 columns and 2 posts. Mr. Jenkins wants to paint the lateral areas of all the columns and posts. How many square feet does Mr. Jenkins have to paint? If Mr. Jenkins can paint about 30 square feet an hour, about how long will it take him to complete the work?

12-7 Skills Practice

Surface Area of Pyramids and Cones

Find the lateral and surface area of each figure. Round to the nearest tenth, if necessary.

1.

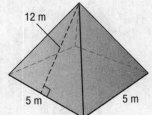

12 m

5 m   5 m

2.

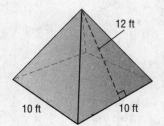

12 ft

10 ft   10 ft

3.

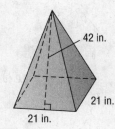

42 in.

21 in.

21 in.

4.

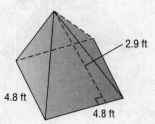

2.9 ft

4.8 ft

4.8 ft

5.

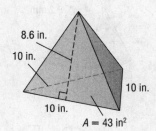

8.6 in.

10 in.

10 in.

10 in.

10 in.

$A = 43$ in^2

6.

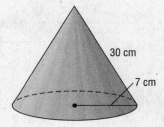

30 cm

7 cm

7.

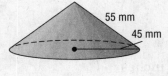

55 mm

45 mm

8.

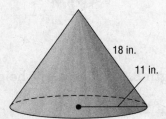

18 in.

11 in.

9.

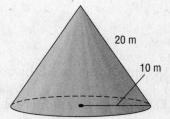

20 m

10 m

10. Square pyramid: base side length 6.3 meters, slant height 4 meters

11. Cone: diameter 16 yards, slant height 10 yards

12. Cone: radius 14 centimeters, slant height 33 centimeters

12-7 Practice

Surface Area of Pyramids and Cones

Find the lateral and surface area of each figure. Round to the nearest tenth, if necessary.

1.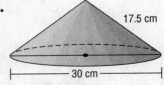

2.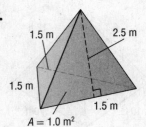

3.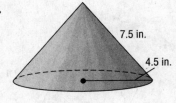

4.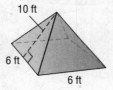

5.

6.

7.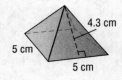

8.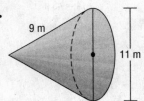

9.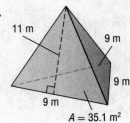

10. Square pyramid: base side length 8.4 inches, slant height 8.4 inches

11. Cone: radius 9 feet, slant height 22 feet

12. Cone: diameter 26 centimeters, slant height 15 centimeters

13. **PAINTING** A wooden structure at a miniature golf course is a square pyramid whose base is 5 feet on each side. The slant height is 4.75 feet. Find the lateral area to be painted.

14. **BAKING** A cone-shaped icicle on a gingerbread house will be dipped in frosting. The icicle is 1 centimeter in diameter and the slant height is 7 centimeters. What is its total surface area?

15. **HISTORY** The Great Pyramid in Egypt was built for the Pharaoh Khufu. The base of each side is 230 meters. The height from the base along the face to the top is 187 meters. Find the total surface area.

12-8 Skills Practice

Similar Solids

Determine whether each pair of solids is similar.

1.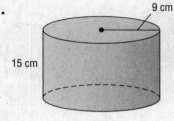

9 cm   3 cm

15 cm   5 cm

2.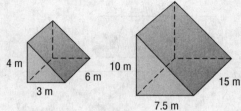

4 m   10 m

3 m   6 m   15 m

7.5 m

3.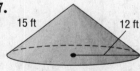

8 in.   3 in.

8 in.   12 in.   3 in.   4 in.

4.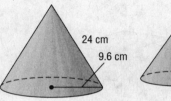

24 cm   20 cm

9.6 cm   8 cm

Find the missing measure for each pair of similar solids.

5.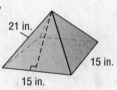

21 in. x

15 in.   21 in.

15 in.   21 in.

6.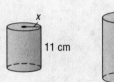

x 6.6 cm

11 cm   16.5 cm

7.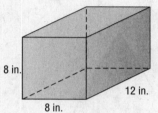

15 ft   12 ft x 8 ft

8.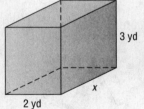

3 yd   2.4 yd

x 1.6 yd   1.8 yd

2 yd

9. A cone has a volume of 600 cubic feet. If the dimensions are reduced by a scale factor of $\frac{1}{2}$, what is the volume of the new cone?

10. A rectangular prism has a surface area of 75 square inches. If the dimensions are tripled, what is the surface area of the new prism?

12-8 Practice

Similar Solids

Determine whether each pair of solids is similar.

1.
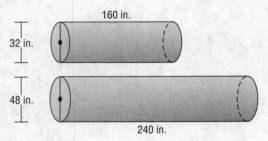

2.
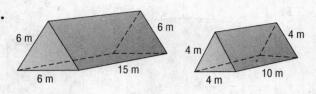

3.
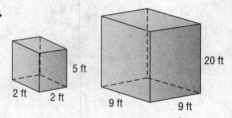

4.
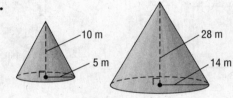

Find the missing measure for each pair of similar solids.

5.
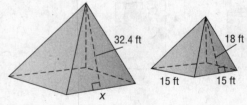

6.
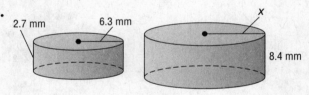

7.
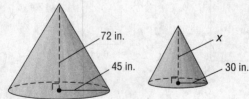

8.
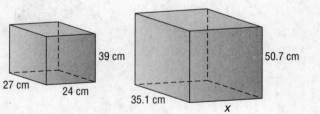

PLAYGROUNDS For Exercises 9 and 10, use the following information.

In the miniature village at the playground, the model of the old school building is 6.6 feet long, 3.3 feet wide, and 4.6 feet high.

9. If the real building was 80 feet long and 40 feet wide, how high was it?

10. What was the volume of the old school building in cubic feet?

13-1 Skills Practice

Measures of Central Tendency

Find the mean, median, and mode for each set of data. If necessary, round to the nearest tenth.

1. 6, 3, 3, 12, 13, 15, 7

2. 1, 1, 0, 2, 1, 1, 0, 0, 1

3. 202, 195, 219, 220

4. 2.5, 4.0, 8.7, 3.3, 3.3, 5.2

5. 21, 23, 39, 44, 27, 25, 28, 30

6. 87, 85, 87, 87, 87

Find the mean, median, and mode for each set of data. If necessary, round to the nearest tenth.

7.

8.

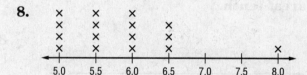

9. TEMPERATURE The average daily temperature by month for one year in Denver, Colorado, is given below. Find the mean, median, and mode for temperature.

Month	Jan	Feb	Mar	Apr	May	June	July	Aug	Sept	Oct	Nov	Dec
Temp. (°F)	43°	47°	51°	61°	71°	82°	88°	86°	78°	67°	52°	46°

Source: *The Universal Almanac*

10. FOOD DRIVE The following set of data shows the number of canned goods collected by each grade at Del Cerro Elementary. Which measure of central tendency best represents the data? Justify your selection and then find the measure of central tendency.

316, 305, 111, 295, 325, 322

13-1   Practice

Measures of Central Tendency

Find the mean, median, and mode for each set of data. If necessary, round to the nearest tenth.

1. 4, 6, 12, 5, 8

2. 16, 18, 15, 16, 21, 16

3. 55, 46, 50, 42, 39

4. 17, 16, 13, 17, 17, 10, 10, 13, 10

5. 25, 25, 25, 20

6. 3.1, 4.5, 4.5, 4.3, 6.0, 3.2

Find the mean, median, and mode for each set of data. If necessary, round to the nearest tenth.

7.

8.

9. TORNADOES   The table below shows the number of tornadoes reported in the United States from 1997–2007. Find the mean, median, and mode for the number of tornadoes. If necessary, round to the nearest tenth.

Year	1997	1998	1999	2000	2001	2002	2003	2004	2005	2006	2007
Number of Tornadoes	1148	1417	1342	1071	1216	941	1367	1819	1264	1106	1074

10. SCHOOLS The following set of data shows the number of students per teacher at different elementary schools in one school district. Which measure of central tendency best represents the data? Justify your selection and then find the measure of central tendency. 13, 15, 11, 15, 20, 14, 16, 16, 13, 17

13-2 Skills Practice

Stem-and-Leaf Plots

Display each set of data in a stem-and-leaf plot.

1. {7, 2, 3, 11, 20, 21, 17, 15, 15, 14}

2. {8, 2, 14, 27, 7, 2, 16, 13, 29, 16}

3.

Amount of Fresh Fruit Consumed per Person in the United States	
Fruit	**Pounds Consumed per Person**
Apples	16
Bananas	27
Cantaloupes	11
Grapefruit	5
Grapes	9
Oranges	11
Peaches and nectarines	5
Pears	3
Pineapples	4
Plums and prunes	1
Strawberries	5
Watermelons	14

Source: U.S. Census Bureau

4.

Winning Scores in College Football Bowl Games, 2007–2008	
Game and Winning School	**Points Scored**
Alamo Bowl, Penn State	24
Cotton Bowl, Missouri	38
Fiesta Bowl, W. Virginia	48
Gator Bowl, Texas Tech	31
Holiday Bowl, Texas Tech	52
Liberty Bowl, Mississippi State	10
New Orleans Bowl, Florida Atlantic	44
Orange Bowl, Kansas	24
Outback Bowl, Tennessee	21
Rose Bowl, USC	49
Sugar Bowl, Georgia	41

Source: ESPN

HUMIDITY For Exercises 5–7, use the information in the back-to-back stem-and-leaf plot. **Source:** The New York Public Library Desk Reference

5. What is the highest morning relative humidity?

6. What is the lowest afternoon relative humidity?

7. Does relative humidity tend to be higher in the morning or afternoon?

U.S. Average Relative Humidity (percent)

Morning		Afternoon
	5	1 2 3 4 7 9
	6	
8 8 4	7	
9 4 0	8	

$8 \mid 7 = 78\%$ $5 \mid 3 = 53\%$

13-2 Practice

Stem-and-Leaf Plots

Display each set of data in a stem-and-leaf plot.

1. {68, 63, 70, 59, 78, 64, 68, 73, 61, 66, 70}

2. {27, 32, 42, 31, 36, 37, 47, 23, 39, 31, 41, 38, 30, 34, 29, 42, 37}

3.

Major League Baseball Leading Pitchers, 2007	
Player and Team	**Wins**
J. Beckett, Boston	20
F. Carmona, Cleveland	19
K. Escobar, Los Angeles	18
J. Lackey, Los Angeles	19
J. Peavy, San Diego	19
C. Sabathia, Cleveland	19
J. Verlander, Detroit	18
C. Wang, New York	19
B. Webb, Arizona	18
C. Zambrano, Chicago	18

Source: ESPN

4.

Average Prices Received by U.S. Farmers	
Commodity	**Price (dollars per 100 pounds)**
Beef Cattle	86
Hogs	49
Lambs	101
Milk	16
Veal Calves	119

Source: U.S. Department of Agriculture

RECREATION For Exercises 5–7, use the information in the back-to-back stem-and-leaf plot shown at the right.

5. The category with the greatest expenditure in 1995 was video and audio goods. What was its total?

6. What is the median expenditure for 1995? For 2005?

7. Compare the total expenditure on recreation in 1995 with that in 2005.

Personal Consumption Expenditure for Recreation (by Category)

1995		2005
8 6 4 4	0	5 6
4	1	0 3
3 1	2	0
	3	8
4 0	4	2
7	5	
	6	7
	7	
	8	2 6

3 | 2 = $23 billion   3 | 8 = $38 billion

13-3 Skills Practice

Measures of Variation

Find the range, interquartile range, and any outliers for each set of data.

1. {7, 9, 21, 8, 13, 19}

2. {33, 34, 27, 40, 38, 35}

3. {37, 29, 42, 33, 31, 36, 40}

4. {87, 72, 104, 94, 85, 71, 80, 98}

5. {92, 89, 124, 114, 98, 118, 115, 106, 101, 149}

6. {6.7, 3.4, 3.8, 4.2, 5.1, 5.8, 6.0, 4.5}

7. {4.3, 1.9, 6.3, 5.1, 2.1, 1.6, 2.4, 5.6, 5.9, 3.5}

8. {127, 58, 49, 101, 104, 98, 189, 111}

9.

Stem	Leaf
1	0 0 3 8 9
2	0 5
3	1 2 4

$2\,|\,0 = 20$

10.

Stem	Leaf
7	8 9
8	1 3 7
9	3 5 6

$9\,|\,3 = 93$

11.

Stem	Leaf
0	2 3 6 8 9
1	2 2 5
2	6
3	2 3 4

$1\,|\,5 = 15$

12.

Stem	Leaf
0	1 1 3 3 7 9
1	2 6 7 8 9 9
2	0 1 2 2 4 5 7 9 9 9
3	2 4 6 7 8
4	0 1 3

$2\,|\,0 = 20$

13.

Stem	Leaf
6	0 6
7	1
8	4 9 9
9	1 3 7 7 7 8

$8\,|\,4 = 84$

14.

Stem	Leaf
4	8
5	1 2 4 7 7
6	0 2 5
7	4

$6\,|\,2 = 62$

HEALTH For Exercises 15–18, use the data in the table showing the calories burned by a 125-pound person.

15. What is the range of the data?

16. What is the interquartile range of the data?

17. Are there any outliers?

18. Which activity burns the most calories per hour? The least calories per hour?

Estimated Calories Burned	
Activity	**Calories Burned per Hour**
Basketball	480
Bicycling	600
Hiking	360
Mowing the Lawn	270
Running	660
Soccer	420
Swimming	600
Weight Training	360
Yoga	240

Source: Fit Resource

NAME _____ DATE _____ PERIOD _____

13-3 Practice

Measures of Variation

Find the range, interquartile range, and any outliers for each set of data.

1. {3, 9, 11, 8, 6, 12, 5, 4}

2. {8, 3, 9, 14, 12, 11, 20, 23, 5, 26}

3. {42, 50, 46, 47, 38, 41}

4. {10.3, 9.8, 10.1, 16.2, 18.0, 11.4, 16.0, 15.8}

5. {107, 82, 93, 112, 120, 95, 98, 56, 109, 110}

6. {106, 103, 112, 109, 115, 118, 113, 108}

7.
Stem	Leaf
1	7 8
2	2 3 5 6 8
3	0

2 | 2 = 22

8.
Stem	Leaf
5	6 7
6	0 1 1 4 8 8 9
7	0 2 3 5 6 7

6 | 1 = 61

9.
Stem	Leaf
4	0 0 0 2 5 7
5	2 6
6	1 8 8
7	0 1 9

5 | 2 = 52

10.
Stem	Leaf
6	4 7 9
7	9
8	1 1 3 3 4 6
9	0 1 2 5

7 | 9 = 79

11.
Stem	Leaf
3	0 1 6 8
4	4
5	2
6	
7	3 3
8	9

5 | 2 = 52

12.
Stem	Leaf
4	3 3 5 7 9
5	0 0 1
6	2
7	4 4 6 8
8	
9	0 1 1 2 2 5

5 | 1 = 51

POPULATION For Exercises 13–15, use the data in the table at the right.

13. What is the range of populations shown?

14. What is the interquartile range for the annual growth rate?

15. Where does the city with the fastest growth rate fall in terms of population? The city with the slowest growth rate?

Populations of the World's Largest Cities 2000		
City	Population millions	Annual Growth Rate (%)
Tokyo, Japan	26.4	0.51
Mexico City, Mexico	18.1	1.81
Mumbai, India	18.1	3.54
Sao Paulo, Brazil	17.8	1.43
New York City, U.S.	16.6	0.37
Lagos, Nigeria	13.4	5.33
Los Angeles, U.S.	13.1	1.15
Calcutta, India	12.9	1.60
Shanghai, China	12.9	−0.35
Buenos Aires, Argentina	12.6	1.14

Source: *World Almanac*

Chapter 13

Glencoe Pre-Algebra

Copyright © Glencoe/McGraw-Hill, a division of The McGraw-Hill Companies, Inc.

13-4 Skills Practice

Box-and-Whisker Plots

Construct a box-and-whisker plot for each set of data.

1. {6, 9, 22, 17, 14, 11, 18, 28, 19, 21, 16, 15, 12, 3}

2. {$45, $37, $50, $53, $61, $95, $46, $40, $48, $62}

3. {14, 9, 1, 16, 20, 17, 18, 11, 15}

4. {$20, $35, $42, $26, $53, $18, $36, $27, $21, $32}

5. {97, 83, 100, 99, 102, 104, 97, 101, 115, 106, 94, 108, 102, 100, 109, 103, 102, 98, 108}

6. {188, 203, 190, 212, 214, 217, 174, 220, 219, 211, 201, 210, 214, 217, 213, 204, 187, 206, 210}

7.

Goals Scored by MLS Leading Scorers		
26	15	12
16	15	1
18	11	10
16	15	5
16	13	9

Source: *World Almanac*

8.

Number of 300 Games per Person in Women's International Bowling Congress		
12	17	23
21	17	23
14	21	12
17	19	24
18	27	13
14	20	12
16	12	16

Source: *World Almanac*

13-4 Practice

Box-and-Whisker Plots

Construct a box-and-whisker plot for each set of data.

1. {14, 30, 35, 8, 29, 28, 31, 42, 20, 36, 32}

2. {$105, $98, $83, $127, $115, $114, $132, $93, $107, $101, $119}

3. {211, 229, 196, 230, 240, 212, 231, 233, 243, 214, 239, 238, 228, 237, 230, 234, 239, 240, 212, 232, 239, 240, 237}

4. {3.7, 6.2, 4.1, 2.4, 1.0, 1.5, 1.4, 2.1, 2.6, 3.0, 1.3, 1.7}

For Exercises 5–7, use the box-and-whisker plot shown.

5. How tall is the highest peak of the Hindu Kush?

6. What is the median height of the major peaks?

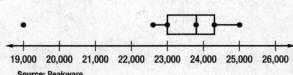

Major Peaks of the Hindu Kush
(height in feet)

19,000 20,000 21,000 22,000 23,000 24,000 25,000 26,000

Source: Peakware

7. Write a sentence describing what the box-and-whisker plot tells about the major peaks of the Hindu Kush.

For Exercises 8–10, use the box-and-whisker plot shown.

8. In which year was the corn yield more varied? Explain.

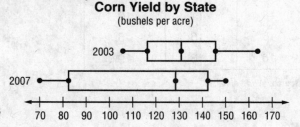

Corn Yield by State
(bushels per acre)

2003

2007

70 80 90 100 110 120 130 140 150 160 170

9. How does the median yield in 2003 compare with the median yield in 2007?

10. Write a few sentences that compare the 2007 yields with the 2003 yields.

13-5 Skills Practice

Histograms

Display each set of data in a histogram.

1.

Shots per Hockey Game		
Number of Shots	Tally	Frequency
1–7	ЖЖ	5
8–14	I	1
15–21	ЖЖ III	8
22–28	II	2
29–35	IIII	4

2.

Employees in Each Office		
Number of Employees	Tally	Frequency
10–19	II	2
20–29	ЖЖ I	6
30–39	ЖЖ IIII	9
40–49	ЖЖ III	8
50–59	I	1

3.

Basketball Backboards on Each Playground		
Number of Backboards	Tally	Frequency
0–4	ЖЖ ЖЖ ЖЖ I	16
5–9	III	3
10–14	ЖЖ III	8
15–19	ЖЖ ЖЖ I	11
20–24		0
25–29	IIII	4

4.

Population of Loons on Local Lakes		
Number of Loons	Tally	Frequency
30–39	II	2
40–49		0
50–59	ЖЖ I	6
60–69	ЖЖ IIII	9
70–79	ЖЖ ЖЖ ЖЖ II	17
80–89	IIII	4

13-5 Practice

Histograms

Display each set of data in a histogram.

1.

Ages of Zoo Volunteers		
Age	Tally	Frequency
18–27	III	3
28–37	IIII III	8
38–47	IIII IIII IIII I	16
48–57	IIII IIII II	12
58–67	IIII	5
68–77	II	2

2.

Crossword Puzzle Solving Times		
Time (min)	Tally	Frequency
0–4	III	3
5–9	I	1
10–14	IIII I	6
15–19	IIII IIII IIII	14
20–24		0
25–29	II	2

For Exercises 3–6, use the histogram at the right.

3. What size are the intervals?

4. How many countries have nine or fewer threatened species?

5. Which interval contains the median number of endangered species?

6. Can you tell from the histogram whether any of the countries have zero threatened species? Explain.

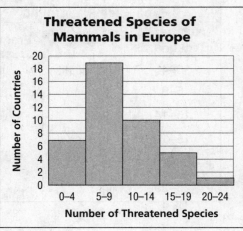

Threatened Species of Mammals in Europe

Source: IUCN

13-6 Skills Practice

Theoretical and Experimental Probability

A spinner like the one shown is used in a game. Determine the theoretical probability of each outcome if the spinner is equally likely to land on each section. Express each theoretical probability as a fraction and as a percent.

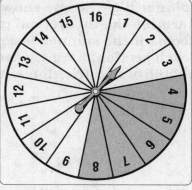

1. $P(10)$

2. $P(\text{odd})$

3. $P(\text{greater than } 7)$

4. $P(\text{prime})$

5. $P(1 \text{ or } 2)$

6. $P(\text{less than } 5)$

7. $P(\text{shaded})$

8. $P(\text{not shaded})$

The table shows the results of an experiment in which the spinner shown above was spun 50 times. Find the experimental probability of each outcome.

Number	Frequency	Number	Frequency
1	II	9	III
2	III	10	HHT
3	II	11	II
4	IIII	12	IIII
5	HHT	13	I
6	HHT I	14	II
7	I	15	III
8	II	16	HHT

9. $P(\text{less than } 7)$

10. $P(\text{even})$

11. $P(\text{not shaded})$

12. $P(13)$

13. $P(\text{greater than } 12)$

14. $P(\text{prime})$

Suppose two 6-sided number cubes are rolled. Find the odds in favor of and the odds against each outcome. (*Hint*: Make a table to show the sample space.)

15. sum of 3 or 5

16. both even numbers

17. odd product

18. sum greater than 10

19. doubles

20. product is a square

13-6 Practice

Theoretical and Experimental Probability

A spinner like the one shown is used in a game. Determine the theoretical probability of each outcome if the spinner is equally likely to land on each section. Express each theoretical probability as a fraction and as a percent.

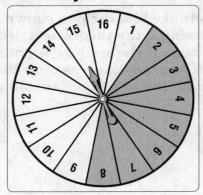

1. $P(15)$

2. P(even)

3. P(greater than 10)

4. P(perfect square)

5. P(shaded)

The table shows the results of an experiment in which the spinner shown above was spun 50 times. Find the experimental probability of each outcome.

6. P(less than 4)

7. P(10 or 11)

8. P(multiple of 4)

9. P(not shaded)

Number	Frequency	Number	Frequency
1	IIII	9	IIII
2	II	10	II
3	III	11	HHI I
4	IIII	12	IIII
5	IIII	13	IIII
6	II	14	II
7	HHI	15	II
8	I	16	I

Suppose two 6-sided number cubes are rolled. Find the odds in favor of and the odds against each outcome. (*Hint*: Make a table to show the sample space.)

10. sum of 6 or 7

11. sum greater than 8

12. sum is a square

The table on the right shows the type and number of businesses in Wilsonville. If there are 625 businesses in the nearby town of Newberry, predict how many of each type of business there would be in Newberry.

13. grocery stores

14. retail stores

15. restaurants

16. pet shops and copy shops

Business Type	Number
Grocery Store	10
Retail Store	54
Copy Shop	6
Restaurant	40
Car Dealership	5
Pet Shop	10

13-7 Skills Practice

Using Sampling to Predict

Identify each sample as *biased* or *unbiased* and describe its type. Explain your reasoning.

1. To determine how many students have pets, all students in one classroom are surveyed.

2. To determine the number of students who plan on attending the Valentine's Day dance, 20 students are randomly selected from each grade level.

3. To determine whether customers are satisfied with their meals, a restaurant collects comment cards that are voluntarily filled out by customers.

4. To determine the most popular color of car, the color of every 12th car that crosses an intersection is recorded.

5. To determine the most popular major league baseball team among its readers, a sports magazine polled a random selection of its readers.

6. **ANALYZE TABLES** The student council would like to sell pizza slices during home basketball games as a fund-raiser. During a home game with 250 people in attendance, they surveyed every 10th spectator to enter the gym about their favorite pizza toppings. Their results are shown in the table. Is this sampling method valid? If so, how many pepperoni pizzas should be ordered if they order 25 pizzas? Explain your reasoning.

Topping	Number
Pepperoni	10
Veggies	8
Cheese	7

13-7 Practice

Using Sampling to Predict

Identify each sample as *biased* or *unbiased* and describe its type. Explain your reasoning.

1. To determine how many people in a town support a new tax levy, 200 people are randomly selected from a phone book and then surveyed over the phone.

2. To determine the number of households in a town that recycle, 40 households from the same street are polled.

3. To determine the usual demand of a Web site, the number of users currently visiting the Web site is recorded every hour.

4. **ANALYZE GRAPHS** The yearbook staff wanted to find out how many students would buy a yearbook. So, the staff surveyed 15 students who were in the school library after school. The results are in the graph. Is this sampling method valid? If so, about how many of the 1287 students in the school will buy yearbooks?

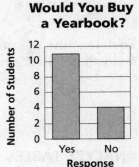

Would You Buy a Yearbook?

5. **LIBRARIES** A library would like to see how many of its patrons would be interested in regularly checking out books from an enlarged print section. They randomly surveyed 200 patrons and 6 patrons responded that they would regularly check out books from an enlarged print section. If the library has a total of 3200 patrons, how many people can they expect to regularly check out books from an enlarged print section?

13-8 Skills Practice

Counting Outcomes

Draw a tree diagram to find the number of outcomes for each situation.

1. Three coins are tossed.

2. A number cube is rolled and a coin is tossed.

Find the total number of outcomes for each situation.

3. One card is drawn from a standard deck of cards.

4. Three six-sided number cubes are rolled.

5. One coin is flipped three consecutive times.

6. One coin is flipped and one eight-sided die is rolled.

7. A sweater comes in 3 sizes and 6 colors.

8. A restaurant offers dinners with a choice each of two salads, six entrees, and five desserts.

Find the probability of each event.

9. Draw the ace of spades from a standard deck of cards.

10. A coin is tossed twice. What is the probability of getting two tails?

11. Draw the six of clubs from a standard deck of cards.

12. Roll a 4 or higher on a six-sided number cube.

13. Roll a 7 or an 8 on an eight-sided die.

14. Roll an even number on an eight-sided die.

15. Draw a club from a standard deck of cards.

16. Roll an odd number on a six-sided number cube.

17. A coin is tossed and an eight-sided die is rolled. What is the probability that the coin lands on tails, and the die lands on a 2?

18. A coin is tossed and a card is drawn from a standard deck of cards. What is the probability of landing on tails and choosing a red card?

13-8 Practice

Counting Outcomes

Find the total number of outcomes for each situation.

1. Joan randomly dials a seven-digit phone number.

2. First-year students at a school must choose one each of 5 English classes, 4 history classes, 5 math classes, and 3 physical education classes.

3. One card each is drawn from four different standard decks of cards.

4. A store offers running shoes with either extra stability or extra cushioning from four different manufacturers.

5. A winter sweater comes in wool or fleece, with a zipper or a crew neck, and in three colors.

6. One spinner can land on red, green, blue, or yellow and another can land on right foot, left foot, right hand, or left hand. Each spinner is spun once.

Find the probability of each event.

7. A number cube is rolled. What is the probability of rolling a four or lower?

8. A number cube is rolled. What is the probability of getting a five or higher?

9. An eight-sided die is rolled and a coin is tossed. What is the probability of landing on an even number and getting heads?

10. A coin is tossed and a card is drawn from a standard deck of cards. What is the probability of landing on heads and choosing a heart?

11. **REFRESHMENTS** How many fruit smoothies are possible from 6 choices of fruit, 4 choices of milk, and 3 sizes?

12. **MONOGRAMS** A school's class rings can include a student's initials in an engraved monogram on the ring. How many different monograms are possible from 2 sizes, 5 type styles, and 3 border styles?

13. **MOBILE PHONES** The table shows the features you can choose for a pay-as-you go phone plan.

   a. How many phone plans have national long distance?

   b. How many customized phone plans include 100 minutes per month talking time and paging capabilities?

Phone	Features	Calling Area	Monthly Talk Time
Brand A; Brand B	e-mail only; paging only; deluxe; paging and e-mail	local only; local and regional; national long distance	30 min; 60 min; 100 min

13-9 Skills Practice

Permutations and Combinations

Tell whether each situation is a *permutation* or *combination*. Then solve.

1. How many ways can 6 student desks be arranged in a row?

2. How many ways can 18 baseball cards be passed out to 2 students?

3. How many ways can 10 students line up for lunch?

4. How many ways can you choose 4 CDs from a stack of 8 CDs?

5. How many ways can 3 pairs of shoes be chosen from 8 pairs?

6. How many ways can 9 runners be arranged on a 4-person relay team?

Find each value.

7. $P(7, 5)$

8. $P(4, 4)$

9. $C(8, 6)$

10. $C(10, 3)$

11. $P(6, 2)$

12. $C(12, 9)$

13. **SPORTS** The Eastern Division of a baseball league is composed of 5 teams. How many different ways can teams of the Eastern Division finish?

14. **LEISURE** The local hobby store has 17 model airplanes to display. If the front case holds 6 models, how many ways can 6 planes be chosen for the front of the store?

15. **ZOOS** The local zoo has 23 animals it can take on visits to schools and other community centers. How many ways can the zoo directors choose 9 animals for a trip to a middle school?

16. **CULTURE** There are 15 Irish dancers in a championship-level competition. How many ways can the top 3 finishers be arranged?

17. **RACING** In an auto race, the cars start in 11 rows of 3. How many ways can the front row be made from the field of 33 race cars?

TELEVISION For Exercises 18 and 19, use the following information.
A television network has a choice of 11 new shows for 4 consecutive time slots.

18. How many ways can four shows be chosen, without considering the age of the viewers or the popularity of the time slots?

19. How many ways can the shows be arranged if the time slots are during prime time and in competition for viewers?

13-9 Practice

Permutations and Combinations

Tell whether each situation is a *permutation* or *combination*. Then solve.

1. How many ways can you make a sandwich by choosing 4 of 10 ingredients?

2. How many ways can 11 photographs be arranged horizontally?

3. How many ways can you buy 2 software titles from a choice of 12?

4. How many ways can a baseball manager make a 9-player batting order from a group of 16 players?

5. How many ways can 30 students be arranged in a 4-student line?

6. How many ways can 3 cookie batches be chosen out of 6 prize-winning batches?

7. **SCHOOL TRIPS** Students are chosen in groups of 6 to tour a local business. How many ways can 6 students be selected from 3 classes totaling 53 students?

8. **CONTESTS** In a raffle, 5 winners get to choose from 5 prizes, starting with the first name drawn. If 87 people entered the raffle, how many ways can the winners be arranged?

9. **RESTAURANTS** A local restaurant specializes in simple and tasty meals.

   a. How many sandwiches are possible if the restaurant lets you build a sandwich by choosing any 4 of 10 sandwich ingredients?

   b. If there are 6 soups to choose from, how many soup-and build-a-sandwich specials are possible?

10. **SPORTS** An inline skate has 4 wheels. How many ways could 4 replacement wheels be chosen from a pack of 10 wheels and fitted to a skate?

GIFT WRAPPING For Exercises 11–14, use the following information.

An upscale department offers its customers free gift wrapping on any day that they spend at least $100. The store offers 5 box sizes (XS, S, M, L, XL), 6 wrapping themes (birthday, wedding, baby girl, baby boy, anniversary, and all-occasion), and 3 styles of bow (classic, modern, and jazzy).

11. How many ways can packages be gift-wrapped at the store?

12. What is the probability that any wrapped package will be in a large box?

13. What is the probability that any wrapped package will *not* have a jazzy bow?

14. What is the probability that a customer will request wrapping for a baby-boy gift?

13-10 Skills Practice

Probability of Compound Events

A number cube is rolled and the spinner is spun. Find each probability.

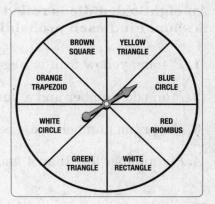

1. P(2 and green triangle)

2. P(an odd number and a circle)

3. P(a prime number and a quadrilateral)

4. P(a number greater than 4 and a parallelogram)

There are 5 yellow marbles, 1 purple marble, 3 green marbles, and 3 red marbles in a bag. Once a marble is drawn, it is replaced. Find the probability of each outcome.

5. a purple then a red marble

6. a red then a green marble

7. two green marbles in a row

8. two red marbles in a row

9. a purple then a green marble

10. a red then a yellow marble

There are 4 yellow marbles, 3 purple marbles, 1 green marble, and 1 white marble in a bag. Once a marble is drawn, it is *not* replaced. Find the probability of each outcome.

11. a purple then a white marble

12. a white then a green marble

13. two purple marbles in a row

14. two yellow marbles in a row

15. a yellow then a purple marble

16. a green then a white marble

A card is drawn from a standard deck of cards. Find the probability of each outcome.

17. P(a red card or a club)

18. P(a diamond or a spade)

19. P(a face card or a 2)

20. P(a 7 or a 9)

21. P(a red card or a king of spades)

22. P(a heart or a queen of diamonds)

13-10 Practice

Probability of Compound Events

An eight-sided die is rolled and the spinner is spun. Find each probability.

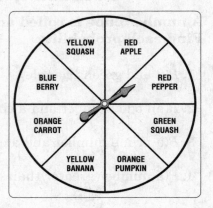

1. *P*(4 and yellow fruit or vegetable)

2. *P*(an odd number and a pumpkin)

3. *P*(a prime number and a red fruit or vegetable)

4. *P*(a number less than 4 and a blue fruit or vegetable)

There are 6 orange marbles, 2 red marbles, 3 white marbles, and 4 green marbles in a bag. Once a marble is drawn, it is replaced. Find the probability of each outcome.

5. a red then a white marble

6. a white then a green marble

7. two orange marbles in a row

8. two marbles in a row that are *not* white

9. a green then a *not* green marble

10. a red then an orange then a green marble

There are 2 green marbles, 7 blue marbles, 3 white marbles, and 4 purple marbles in a bag. Once a marble is drawn, it is *not* replaced. Find the probability of each outcome.

11. a green then a white marble

12. a blue then a purple marble

13. two blue marbles in a row

14. two marbles in a row that are *not* purple

15. a white then a purple marble

16. three purple marbles in a row

The chart shows the letter-number combinations for bingo. The balls are randomly drawn one at a time. Balls are *not* replaced after they are drawn. Find the probability of each outcome.

B	I	N	G	O
1	13	25	37	49
2	14	26	38	50
3	15	27	39	51
4	16	28	40	52
5	17	29	41	53
6	18	30	42	54
7	19	31	43	55
8	20	32	44	56
9	21	33	45	57
10	22	34	46	58
11	23	35	47	59
12	24	36	48	60

17. a B-1

18. a G

19. an N or a B-2

20. an I or an O

21. *not* a G

22. a B-6, then a G, then another G